U.S.–ASIAN RELATIONS

U.S.–ASIAN RELATIONS

The National Security Paradox

Edited by
James C. Hsiung

PRAEGER SPECIAL STUDIES • PRAEGER SCIENTIFIC

New York • Philadelphia • Eastbourne, UK
Toronto • Hong Kong • Tokyo • Sydney

Library of Congress Cataloging in Publication Data

Main entry under title:

U.S.-Asian relations.

  Bibliography: p.
  Includes index.
  Contents: Asia and Reagan's global strategic design / James C. Hsiung— In search of a strategy, the Reagan administration and security in Northeast Asia / Norman D. Levin—U.S.-China military ties, implications for the U.S. / Robert G. Sutter—[etc.]
  1. United States – Military policy–Addresses, essays, lectures. 2. United States–Military relations–Asia–Addresses, essays, lectures. 3. Asia–Military relations–United States–Addresses, essays, lectures. 4. United States–National security–Addresses, essays, lectures. 5. United States–Foreign relations–1981–Addresses, essays, lectures. I. Hsiung, James Chieh, 1935-   . II. Title: U.S.-Asian relations.
UA23.U15   1983   355'.0335'73   83-13930
ISBN 0-03-064189-6

Published under the Auspices of the Contemporary U.S.-Asia Research Institute, Inc., New York, New York.

Published in 1983 by Praeger Publishers
CBS Educational and Professional Publishing
a Division of CBS Inc.
521 Fifth Avenue, New York, NY 10175 USA

©1983 Praeger Publishers

All rights reserved

3456789 052 987654321

Printed in the United States of America
on acid-free paper

Preface

The United States in the early 1980s is faced with new challenges and opportunities in the foreign policy area. The new challenges came with the uneasy knowledge that the Soviet Union has caught up with the United States in military might, not the least of which is nuclear weaponry. The Soviet military presence in Afghanistan and Indochina, the rumblings of renewed Soviet activism in the Middle East following the 1982 Israeli invasion of Lebanon, and massive expansion of the Soviet Pacific Fleet are but acute indications of the Soviet threat, as perceived from Washington. At the same time, these signs of Soviet expansionism also help put the limelight on larger Asia, the region under study in the present volume. The new opportunities, on the other hand, were presented by the succession of Yuri Andropov in the Kremlin, following 18 years of the Brezhnev era.

Ronald Reagan, as candidate and as President, has staked out a program of "rearming America" and "making America strong again." Increasing the United States' defense outlay for strategic and conventional forces, to the tune of $1.76 trillion for 1984–1988, is billed as a necessary means to arrest the perceived U.S. power decline since the Vietnam War. That seems to take precedence over strategic arms control negotiations with the Soviet Union, although the troubled START* finally began in the summer of 1982. The Administration also advocates a *global defense* strategy, to confront the Soviets in three areas: Western Europe, the Persian Gulf, and the Indian–Pacific Oceans.

The strategy entails, among other things, the building of a "three-ocean navy," the deployment of the MX missile, the production of the B-1 bomber, the development of the Trident II SLBM (submarine-launched ballistic missile), and the revival of the ABM (anti-ballistic missile system) option, as well as proceeding with the plan, initiated under President Carter, to install 572 Pershing II and ground-launched cruise missiles in Western Europe by the end of 1983. The plan also calls for the strengthening of U.S. conventional forces to be able to fight a *sustained* subnuclear war, for months, rather than for weeks.

The kind of tough talk generated by the Reagan team appears likely to fan the fears of nuclear war, instead of allaying them, both at home and

---

* START, or Strategic Arms Reduction Talks, has replaced SALT (Strategic Arms Limitation Talks), a process begun with the Nixon Administration.

abroad (especially in Western Europe). Often it sounds as though the sole function of foreign policy is to combat the Soviet Union throughout the globe. Critics of the Reagan plan have characterized it as providing a mere weapons "wish list" without any clear priorities, nor a coherent strategy on how to project U.S. power in world affairs. In fact, it has often been said that the Administration has no coherent foreign policy, let alone an Asian policy, other than tough talk, persistent defense budget hikes, and high-sounding goals. In this volume, we hope to find an answer to the following questions: (1) Is there a Reagan foreign *policy*, particularly with respect to Asia? (2) If so, *what* is that policy? and (3) How, or by what *strategy*, is that policy translated into action? and with what *results*? To the extent possible, we hope also to ascertain how the Reagan policy is different from that of previous administrations.

Because of the avowed emphasis of the Reagan Administration on the security and strategic aspects of foreign policy, we hope, therefore, to examine those aspects more closely. Hence, the title of this volume is deliberately chosen to reflect that focus. The purpose of this work is evident in the arrangement of the chapters, beginning with an introductory chapter exploring systematically the administration's global security plan and how Asia fits into it, followed by individual chapters looking into the subregional units or countries, and capped by a concluding chapter that takes another overview in retrospect, drawing on the findings by the various contributors. Asia, in this volume, is used broadly and loosely, to include that part of the map that stretches from Northeast Asia (Korea, Japan, China, and the Soviet Far East), Southeast Asia (Indochina, which consists of Vietnam, Laos, and Kampuchea; plus the ASEAN nations, i.e., Thailand, Malaysia, Singapore, Indonesia, and the Philippines), through the western Pacific and the Indian Ocean, to the Persian Gulf. Although the chapters do not follow the map so neatly and literally, they together cover the geographical span as suggested. The beginning and concluding chapters attempt to fill the gaps and above all to place Asia, so defined, into the global picture.

Our critics, I know, will find the book either too ambitious (because of the vast extent it covers) or insufficient (since South Asia is not exam-examined, and Southwest Asia is only tangentially treated within the context of the Persian Gulf and the Reagan Middle East peace plan). My defense, as editor and architect, is that this is not a survey volume dutifully covering each and every component part of the region. It is, instead, mainly an inquiry into the effects of global security management on the conduct of U.S. foreign relations in Asia, more especially under Reagan. Moreover, although as indicated in my introductory chapter it is imperative to apply a global perspective to the inquiry, we have to draw a line somewhere. We have to be broad enough to be able to ascertain the interrelations between central strategy and regional operations, and yet be focused enough as not to go astray from our central themes.

This volume is under the auspices of the Contemporary U.S.–Asia Research Institute, Inc., a private nonprofit research organization dedicated to the study of U.S.–Asian and intra-Asian relations. Contributors to the book, nevertheless, wrote from their conscience and do not purport to represent the views of the Institute. Part of my own research related to the subject was undertaken during my sabbatical in 1982, which took me to a number of Asian countries. Many people have helped to make this volume possible. In addition to the contributors, I am thankful to Professor Shao-chuan Leng, of the University of Virginia (Charlottesville), Dr. Arthur Lim and Director K. S. Sandhu, of the Institute of Southeast Asian Studies (Singapore), General Cheng-chung Li (retired), of the Society for Strategic Studies, Republic of China (Taipei), Dean Benjamin Young, of the Chung Ang University (Seoul), Dr. Lillian Craig Harris, of the U.S. Department of State, Dr. Martin Sours, of the American Graduate School of International Management (Phoenix, Arizona), Professor Yuan-li Wu, of the Hoover Institution, Ms. Carol Wang, of the C.W. Enterprises and Investment, Inc. (San Francisco), among others, for their support, assistance, and/or encouragement. My thanks also go to my assistants at New York University: Patricia Lane, Lauren Lung, and more especially Omer Karasapan, who did much of the legwork and compiled the bibliography. Deborah Bula and Cynthia C. Hsiung, my daughter, typed parts of the manuscript. While the contributors are responsible for the views presented in their own chapters, I alone bear the responsibility for the overall quality and design of the product.

J.C.H.

Contents

Preface   v

Chapter

1. Asia and Reagan's Global Strategic Design   1
   James C. Hsiung

2. In Search of a Strategy: The Reagan Administration and Security in Northeast Asia   19
   Norman D. Levin

3. U.S.–China Military Ties: Implications for the United States   33
   Robert G. Sutter

4. Emergence of an "Independent" Chinese Foreign Policy and Shifts in Sino–U.S. Relations   63
   Carol Lee Hamrin

5. The Reagan Administration's Southeast Asian Policy   85
   John W. Garver

6. In Search of Peace in the Middle East   131
   Winberg Chai

7. Invisible Enemies: Conflict and Transition in Soviet–U.S. Relations   147
   Paul H. Borsuk

8. National Security and the Policy Paradox   169
   James C. Hsiung

Bibliography   191
Index   203
About the Editor and Contributors   209

1

Asia and Reagan's Global Strategic Design

James C. Hsiung

The Reagan Administration's Asian policy both mirrors its larger global strategic concerns and follows a traditional U.S. reflex toward Asia. The global concerns stem from a perceived threat of an expanding Soviet power, which is no longer land-based and confined to Europe, but truly intercontinental. In response to Moscow's expanding nuclear might and the projection of its naval power to the Indian and Pacific oceans, Washington has adopted a strategy of confronting the Soviets on three fronts: NATO* to the west, the Asian–Pacific region to the east, and the Persian Gulf to the south. The dictate of combating the Soviet threat in Asia, from the Western Pacific to the Persian Gulf, has converted the region into a locus of direct U.S.–Soviet competition. This increased acuteness in the region's strategic importance reinforces a traditional U.S. policy tenet with regard to Asia, which (1) conceives of the United States as a Pacific power, no less than it is an Atlantic power, and (2) requires that the Asian–Pacific region not be dominated by any unfriendly nation or coalition of nations. All three wars the United States fought in the region during the present century—World War II (with Japan), the Korean War (against North Korea in league with Communist China), and the Vietnam war (against an assumed international Communist conspiracy)— were related to U.S. fears of such a regional domination by a hostile power or bloc. Because of this traditional reflex, the rise in Soviet military presence in the area is considered all the more ominous, especially by a conservative President who feels strongly about reasserting U.S. power abroad to counter the military and ideological menace posed by Moscow.

---

* To be more exact, Western Europe and the North Atlantic community within the perimeter of the North Atlantic Treaty Organization (NATO).

Thus, President Reagan's Asian policy is inseparably intertwined with his global campaign to constrain Soviet influence. It is necessary, therefore, for us to dwell on his global strategic policy to provide the backdrop for the discussions on his Asian policy in the chapters that follow.

THE PERCEPTION OF SOVIET EXPANSIONISM

In the 37 years of U.S.–Soviet conflict since the end of World War II, every U.S. President, Democratic or Republican, has felt a Soviet threat to U.S. security. But, unlike Presidents Nixon (who was more preoccupied with the maintenance of a favorable balance of power vis-a-vis the Soviet Union) and Carter (who focused more on Soviet human rights violations, at least until Afghanistan), President Reagan's abhorrence of the Soviet Union is directed at its threat to the very existence of the Western free enterprise system and the values the latter espouses. He displays a devout sense of mission in the safeguarding of that system against the Soviet challenge. Ths sense extends to Reagan's defense of the Western marketplace values against the encroachments by other socialist ideologies, such as has been demonstrated in his hostility toward the Law of the Sea Treaty concluded in 1982. The treaty's main anathema is its requirement that future exploration of seabed resources be licensed by an International Seabed Authority (ISA) to be established under United Nations auspices. The "common legacy of mankind" principle that underscores the ISA's licensing authority is offensive to Reagan's strong conviction in the private enterprise philosophy.[1] The same concern for the sanctity of capitalist institutions had led the President, however paradoxically, to keep General Jaruzelski's Poland from defaulting on its international loan interest payments, shortly after martial law was installed, as Reagan intervened to protect the integrity of the Western banking system.[2]

If Reagan's anti-Soviet hard line is reminiscent of much of the Cold War rhetoric, he has to cope with the new reality of the 1980s that no President, from Truman to Kennedy, ever experienced. The United States no longer enjoys the nuclear superiority with which John F. Kennedy, during the 1962 missile confrontation, could overawe Khrushchev to blink first, not to mention the nuclear monopoly during Harry Truman's presidency.[3] The nuclear gap now perceived in Soviet favor is not just a matter of defense for the Reagan Administration. In the context of the ideological competition between the two camps, it seems to impute a superiority to the very system that has produced the alleged nuclear superiority. Furthermore, in the perception of the Republican conservatives surrounding the President, Moscow has gained its nuclear edge in recent years only as a result of the lopsided benefits that detente has bestowed on the Soviets. At his first Presidential press conference, on January 29, 1981, Reagan defined detente as a "one-way street that the Soviet Union has used to pursue its own aims."[4] More than Jimmy Carter, President Reagan not only sees that the current nuclear balance is more favorable to Moscow, but believes that the nuclear advantages might encourage Moscow to risk confrontations.[5]

This concern about a "widening gap" between the U.S. and Soviet nuclear arsenals went back to candidate Reagan's 1980 Presidential campaign.[6] It received more explicit articulation during the President's news conference on March 31, 1982:

> On balance, the Soviet Union does have a definite margin of superiority, enough so that there is risk and there is. . .a window of vulnerability.[7]

Beneath the shallow surface of the "window" metaphor are two underlying themes that support the President's thinking: (1) The Soviet ICBMs (intercontinental ballistic missiles) pose the threat of destroying all U.S. land-based missiles in one sweep, whereas the latter pose no corresponding threat to Moscow's; and (2) NATO labors under a similar disadvantage, in that every military target in Western Europe is within the range of SS-20 ballistic missiles based in the Soviet Union, while NATO now has no similar weapon that returns the courtesy.[8]

This perception of vulnerability is responsible for three interrelated strategic elements that characterize Reagan's national security policy: (1) a determination to redress the imbalances in nuclear and conventional forces; (2) a global defense against the real or potential onslaught of the Soviet Union and its cohorts, including their alleged "terrorist" acts; and (3) a strategic doctrine that support both (1) and (2). We discuss these three elements separately.

TO REDRESS THE NUCLEAR AND CONVENTIONAL IMBALANCES

The Reagan Administration has acted in several ways in an attempt to close the perceived gap, ranging from a redoubled effort to strengthen the nation's warfighting capability (including both nuclear and conventional forces, at both strategic and theater levels), an ambitious civil defense program, to a new approach to nuclear arms reduction.

Closing the Nuclear Window

The Reagan team in early 1982 put forward a comprehensive $1.6 trillion, 5-year defense program, covering fiscal years 1983–1987.* Although only 5% of the proposed budget was earmarked for nuclear modernization,[9] the Administration is seeking, through a two-pronged strategy of unilateral buildup and bilateral nuclear reduction talks with Moscow, to close the nuclear gap. In addition to beginning the production of the B-1 bomber (scratched by Carter in 1977), Reagan decided in November 1982 to embark on a $26 billion scheme to deploy the new MX supermissile. Although its proposed "dense pack" basing mode, which replaced the mobile "shelf-game" version endorsed by Carter, was immediately hooted down in Congress, the MX missile decision demonstrated President Reagan's resolve to remedy the vulnerability of U.S. ICBMs by making it impossible for Soviet missiles to wipe out all U.S. land-based missiles in one strike.[10] The MX, which was first conceived during the Nixon Administration, is designed to destroy even the most heavily armored Soviet silos; hence, a direct answer to the threat posed by the Soviet SS-18 intercontinental supermissiles.

Under the Reagan plan, a nuclear warhead larger than that of the Mark 12 A's now carried by the Minuteman II missile is being developed for the MX. To strengthen U.S. first-strike capability, the Trident II submarine-launched missile force (SLBMs) will also be expanded. In addition to improved accuracy, the Trident II will, going beyond the Carter plan, have an improved range as well.[11]

Another Reagan initiative is the revival of the ABM (anti-ballistic missile system) option, which was on the backburner following the 1973 U.S.–Soviet agreement reached by Presidents Ford and Brezhnev, that modified the 1972 ABM Treaty. Both sides feared then that the installation of the ABMs would be destabilizing and hence self-defeating, because it would destroy the MAD (mutual assured destruction) prerequisite for nuclear deterrence to work. The Reagan team, however, considered ABMs necessary for the survival of the MX missiles. As part of its military budget for fiscal 1983, the Pentagon recommended a doubling of spending by the Army's ABM research program, to $727 million. Subject to Congressional approval, the Pentagon sought to spend $1.5 billion more per year, from fiscal 1984 through 1989, on research and development for the Sentry low-altitude ABM. Procurement would cost a further $9 billion to $12 billion. The planned deployment of an ABM system would eventually require renegotiation of the ABM treaty with Moscow.[12]

---

* In January 1983 the Pentagon put forward a new $1.76 trillion defense budget, covering 1984–1988. *New York Times*, February 1, 1983, p. A17.

The Administration also sought the production of a new version of the B-1 bomber, called the B-1B, the continuing development of the "stealth" bomber, designed to evade enemy radar detection, and the modernization of existing B-52 bombers. It also wanted to continue developing the 1,500-mile-range ALCM (air-launched cruise missile) and to produce enough of them to be carried by the B-52's and eventually by B-1B's, to strike targets from beyond the reach of Soviet air defense.[13]

Civil Defense Program

Reagan's conception of deterrent credibility relies more heavily than Carter's on defending the United States proper, as well as being able to attack the Soviet Union. In a clear departure from Carter policy, the Reagan team designed an ambitious 7-year, $4.2 billion civil defense program that, it claims, would allow evacuation of most large U.S. cities in case of a severe international crisis. Despite severe criticisms from many quarters, it believed that the civil defense program, which would theoretically protect 90% of the U.S. population against the effects of a Soviet nuclear attack, will contribute to deterrence by making a resort to war a more credible U.S. option.

Nuclear Reduction Talks (START)

Besides unilateral nuclear modernization, the other prong in Reagan's strategy to close the nuclear window is to negotiate a reduction of the existing ballistic warheads and throw weight. This, however, came as a secondary measure to the United States's own nuclear rearmament outlined above. The President did not announce a new proposal to achieve reduction until May 9, 1982, almost 1½ years after he took office. The negotiations, which began on June 29, are called START (Strategic Arms Reduction Talks), to dramatize the goal of nuclear weapons reductions and to signify a break from the previous SALT (Strategic Arms Limitation Talks) approach and the unratified SALT II treaty.[14]

To the Reagan Administration, the fault of the SALT approach was that it gave one-sided advantages to the Soviets. For example, the limitation it sought was based merely on the counting of launchers, which "would equate a Soviet SS-18 carrying 10 warheads with a U.S. Minuteman II carrying a single warhead."[15] The START proposal would seek, in the first place, to achieve a significant reduction in ballistic missile warheads (to about 5,000 for each side, no more than half of which would be deployed on ICBM's) and in ballistic missiles (to 850 on each side). In the second phase, further reductions to equal ceilings would be sought on other elements of strategic forces, particularly missile throw weight. Throw weight is an important measure of the size and destructive potential of ballistic missiles. First-phase reductions are aimed at reducing the current disparity in ballistic missile throw weight and laying the groundwork for the second-phase reductions to achieve an equal throw weight ceiling below current U.S. levels.[16]

Parallel to START, which deals with strategic (or global) nuclear weapons, was the President's "zero option" proposal to deal with intermediate-range (or theater) missiles, which he announced in late 1981. Under this plan, the United States would cancel its projected deployment of Pershing II and ground-launched cruise missiles if the Soviets would dismantle their existing SS-20, SS-4, and SS-5 missiles. [17]

Conventional Balance

As the allocations of the proposed 5-year $1.6 trillion defense budget indicate, nuclear deterrence, however important it may be, is not the only pillar of U.S. security. Reaganite strategists have embraced the premise that credible deterrence now requires that the United States and its allies be prepared to fight a longer conventional war with the Soviet Union—longer than just a brief conventional "pause" that serves as a trip wire for nuclear escalation. Underlying the importance of the conventional deterrence scenario is the fact that nuclear arms almost never could be used since the opponent could respond in kind. The United States, it was argued, must be able to match the Soviets in conventional forces so that the latter would not be able to use its conventional preponderance to intimidate Washington.[18]

The same dual-track approach noted above applies to the management of the conventional balance with the Soviet rival. On the one hand, the Administration initiated a new proposal calling for a single comprehensive agreement for the MBFR (mutual and balanced force reductions) talks. Ten NATO members and seven Warsaw pact nations are involved in the talks. Under the agreement, all direct participants with major military formations in Central Europe would undertake to reduce ground forces to a common collective ceiling on each side of about 700,000 ground personnel and about 900,000 ground and air force personnel combined. Reductions, according to the proposed agreement, would be accomplished in four verifiable stages within 7 years.[19]

On the other hand, the Administration was also endeavoring to upgrade the quality and levels of U.S. conventional forces. The centerpiece of the plan is a "600-ship Navy," as opposed to the 479 ships the Navy had when President Carter left office. The 600-ship target was derived from a fundamental decision to deploy 15 aircraft carriers (up from 13), four missile-armed battleships, and enough amphibious ships to carry a

Marine division and one additional brigade (about one-third of a division). The last two categories will be new to the Navy, which now has no battleships and barely enough amphibious ships for a Marine division. The principal change planned for Air Force combat units is an expansion from 36 fighter wings (24 active and 12 reserve wings). Although small units would be added to the existing Army divisions, the Reagan team planned no expansion of the current force of 16 active and 8 reserve divisions until 1986 or 1987, when two more reserve divisions will be organized. Compared to the Carter program, the Reagan plan "will provide ground forces with 29% more M-1 tanks, 34% more fighting vehicles, and 25% more [anti-tank] attack helicopters," according to Under Secretary of Defense Fred Iklé.[20]

A GLOBAL DEFENSE

Like every Administration since World War II, the Reagan government wants to shield West Europe and Japan (and South Korea as well) from Soviet coercion. In a way, this definition of vital U.S. interests is in keeping with the concept of the "Rimland" à la Spykman,[21] according to which the United States must defend the east and west rims of the Eurasian land mass—namely, West Europe and the Chinese subcontinent—against the threat of the "Heartland" power (i.e., Russia). Due to the temporary "loss" of the Chinese mainland from the U.S. strategic map, 1949-1972, the idea of the rimlands has been extended, on the east, to Japan—just as it has reached out, on the west, to Africa. The extension in scope has survived China's "return" following Nixon's historic trip of 1972 and the normalization of its relations with the United States under President Carter. With the Reagan Administration, there seems to be a further geographical extension of the rimland concept to include the Indian and Pacific oceans and the Persian Gulf, both because of the resources these areas command and also the unprecedented projection of Soviet naval power.[22]

In recent years the term "global," with reference to U.S. security strategy, increasingly has come to encompass not just the continents, but the oceans as well. If the Atlantic used to be the only ocean that counted in U.S. defense concerns—dubbed by Sir Halford Mackinder as the "midland ocean," to suggest the enclosed proximity between the shores of the North Atlantic[23]—this is less true today. In the current usage in Washington, "global" defense means stretching security concerns from the NATO community to the Middle East/Persian Gulf, to the Indian and Pacific oceans, and to the Caribbean and the Americas as a whole. The commitment to Europe remains the central premise of U.S. policy as before. Simultaneous attention is directed to the projection of Soviet power elsewhere, especially where the security of U.S. major allies is concerned, including Japan and South Korea.

The Security of Western Europe

The most distinct reminder of the U.S. commitment to Western Europe is the continuing presence of five U.S. Army divisions and 30 Air Force squadrons. The Reagan team's goal is a capability to fly another five U.S. divisions and 60 fighter squadrons, of about 24 planes each, to Europe within 10 days. In a clear break from past administrations, there are plans, according to John F. Lehman, Jr., Navy Secretary, to amass several carriers and a Marine-laden amphibious force in the Norwegian Sea, at the outset of hostilities, to seal off the routes by which Soviet planes and submarines based in the Arctic would attack transatlantic supply routes. Carter officials had discounted the scenario as impractical because a naval force would not survive being so close to Soviet bases that early in a war.[24] Reagan has reaffirmed the Carter plan to deploy, starting in late 1983, 572 Pershing II missiles and ground-launched cruise missiles in Western Europe. These missiles, which will place Soviet targets within reach from Western Europe, will greatly alter the military balance there in countering a Soviet force of at least 330 triple-warheaded SS-20 missiles (out of a total of 600 assorted missiles), which can reach any part of Europe from the Ural Mountains.

Defending the Persian Gulf

As demonstrated in the 1973 oil embargo, access for the United States and its key allies to economically vital raw materials is of strategic importance. These materials include petroleum from the Persian Gulf region and various metals and other minerals from southern Africa. Oil from the Persian Gulf accounted for 30% of U.S. petroleum consumption in 1981. Europe and Japan imported 60 and 70%, respectively, of their oil from the Persian Gulf during the same year. Since the Carter years, the United States has considered its role in the defense of the region as absolutely vital for securing the free world's access to its oil supplies. The Reagan team insists on preparing to counter any Soviet attempt to seize the gulf region by force, even though such a move is not likely in the foreseeable future, at least less likely than seizure by Khomeini's Iran or a radical leftist state.[25]

Continuing the basic Carter blueprint, the Reagan plan involves: (1) the maintenance of at least one (sometimes two), aircraft carrier in the Indian ocean at all times, and (2) the readiness to send rapid deployment force (RDF) units to the gulf region at the invitation of an area state. Forces earmarked for the RDF include three Army divisions, two Marine divisions, and eleven Air Force fighter wings. Although not directly linked to the Gulf, one Reagan initiative is his commitment to enlarge the fleet of amphibious landings from which Marine units would shoot their way ashore in the face of hostile combat forces. This could expand U.S. options in the Persian Gulf region. In addition to the "landing only if invited" scenario, the emphasis on the Marines' "forcible entry" capacity signifies a preparation for U.S. seizure of Gulf oil fields in case of local domestic turmoil.

The Indian and Pacific Oceans

The crises in Iran and Afghanistan have necessitated the strengthening of U.S. naval forces in the northwest Indian Ocean, considerably stretching the area patrolled by the Seventh Fleet. The one aircraft carrier regularly maintained in the Indian Ocean, mentioned above, is assigned from the Sixth Fleet in the Mediterranean to the Seventh Fleet. The latter used to have duties in the waters of East and Southeast Asia (the western Pacific) only.[26] The most important lesson of the recent southwest Asian crises is that this part of the world is no longer an appendage of the European theater, as it was after World War II. Southwest Asia is now becoming linked to the Asia–Pacific theater, just as the Indian Ocean is linked to the western Pacific Ocean. A most dramatic indication of this linkage is the U.S. consciousness that defense of Southwest Asia contributes directly to the defense of Japan, on account of the latter's overwhelming dependency on Middle East (especially Persian Gulf) oil.[27]

Washington now emphasizes the importance in each area of being able to protect seaborne supply routes against Soviet or other forces. The continuing presence of the one carrier in the Indian ocean, though much of it for its presumed psychological or symbolic significance, is to correct the erstwhile undesirable "swing" strategy. Under that strategy, during the Iranian and Afghan crises of 1980, two Seventh Fleet carriers were dispatched to the area, leaving no carriers in East and Southeast Asia, creating political repercussions in Japan and Korea.[28]

In the western Pacific, the Navy now maintains that it no longer can plan to "swing" carriers from this region to the Atlantic in case of a crisis, for another reason. The recently expanded Soviet Pacific Fleet, which has duties in both the western Pacific and the Indian oceans, had a total of 720 warships in 1981. These include one V/STOL aircraft carrier (the 40,000-ton *Minsk*), 85 major surface ships, and 130 submarines (57 of them nuclear-powered). Detachments from the Pacific Fleet (average 3 submarines, 7 surface combatants, and 18 support ships) serve in the Indian Ocean. The U.S. Seventh Fleet, on the other hand, had 47 warships, including 3 carriers, 21 major surface combatants, and 8 nuclear-powered submarines. Detachments from the Seventh Fleet serve in both the Indian Ocean (1 carrier and 6 surface combatants) and the Persian Gulf (1 command ship and 4 surface combatants). The Soviet Pacific Fleet had 420 aircraft, as against the U.S. Seventh Fleet's 262 seaborne aircraft. With 1.5 million tons of ships, the Soviet fleet more than doubled its U.S. counterpart in total tonnage.[29]

While the U.S. side is qualitatively ahead and its total tonnage should include the naval strength of the Japanese, the challenge posed by the expanding Soviet fleet is formidable.[30] The Reagan team was arguing that the U.S. Seventh Fleet must be large enough to protect Japan, South Korea, and Pacific sea lanes by attacking the complex of air and naval bases near Vladivostok and Petropavlosk. The aim of creating a "600-ship navy" is in part to counter the Soviet naval threat in the Western Pacific and Indian oceans, where nearly 100,000 tons of ships are added annually to the Soviet Pacific Fleet. There appears to be a growing awareness in Washington of the complementarity between the Soviet SLBM sanctuary in the Okhotsk Sea (in the east) and that in the Barents Sea and adjacent Arctic waters (in the north). Although 70% of Moscow's SLBM fleet is still harbored in the north, and the Far Eastern locale is being used as a fallback in the early 1980s, the increasing congestion of more Delta-class ships (armed with intercontinental missiles) in the Barents and Arctic waters has made the Okhotsk sea increasingly more valuable for the Soviet navy.[31] These developments are bound to pose questions for U.S. priorities with regard to the western and northern Pacific as compared with other areas.

To round out the *global* picture in the Reagan defense plan, it is instructive to take heed of its view toward the Caribbean. Whereas Carter recognized Cuba as a military problem, following the 1979 discovery of a Soviet Army brigade in that country, it was nevertheless still considered merely a thorn in Uncle Sam's thigh, that is, a localized threat to the sea lanes to and from U.S. Gulf Coast ports, or a real or potential supplier of men and materiél to left-wing revolutionaries in Latin America. Reagan's defense aides have greatly accentuated the theme of Cuba in the context of the relevance of the Caribbean to NATO's security. "In peacetime," as Secretary Weinberger put it, "44 percent of the crude oil imported into the United States pass through the Caribbean. In wartime, half of NATO's supplies would transit by sea from Gulf ports through the Florida Straits."[32] No region, in this light, is to be compartmentalized from the rest of the world, or, in other words, is considered irrelevant to the anti-Soviet cause.[33]

THE UNDERLYING STRATEGIC DOCTRINE

It is pertinent to ask whether all that the Reagan Administration proposes to do in defense, as outlined above, has any coherent strategic doctrine beyond the obvious unifying anti-Soviet theme. There is, unfortunately, no one spokesman for the Administration on this question, and there is no one central document which will provide an answer. We therefore have to attempt to deduce an answer from both Reagan's defense program and the public statements made by the President and other key officials. In doing so, we shall try to bring some coherence to the various ingredients one finds, especially where coherence is either unclear or wanting.

Peace through Strength

The first generalizable theme is "peace through strength," a slogan obviously coopted from conservative groups such as the Committee of the Present Danger, the American Security Council, and so on, which have repeated the line for sometime. During his campaign, candidate Reagan echoed the theme. "The burden of maintaining peace," he declared in his October 29, 1980, televised debate with Carter, "falls on us. And to maintain the peace requires *strength*. America has never gotten in a war because we were too strong."[34] The Reagan Administration's program for modernizing nuclear and conventional forces, already outlined, was precisely designed to increase the country's "strength" in the name of peace. The same theme has been emphasized by various Reagan aides on different occasions. "The United States," declared Secretary Weinberger, "will maintain a strategic nuclear force posture such that, in a crisis, the Soviets will have no incentive to initiate a nuclear attack on the United States or our allies."[35]

Secretary of State Alexander Haig, Jr., used to characterize "America's economic and military *strength*" as the "first pillar of our foreign policy."[36] Secretary Shultz, his successor, continued the same party line, when he stated "America's yearning for peace does not lead us to be hesitant in developing our *strength* or in using it when necessary." Shultz then defined "military power for peace" and "a sound economy" as the "bulwark" and "engine," respectively, of U.S. strength.[37] In an address on "President Reagan's Framework for Peace," William Clark, National Security Adviser to the President, declared that next to economic recovery at home, the President's "second essential task to keeping the peace was to restore the foundation of *strength*, which underwrites the concept of deterrence."[38]

If anything, the argument that more strength spells greater chances for peace sounds very much like an equivalent of "supply side economics" in deterrence. Its two practical assumptions are that (1) more military hardware in our hands will overawe the Soviets into behaving responsibly (hence, deterrence), and (2) readiness to use our might, including first use, is the only sure way to check Soviet expansionism. As throughout this chapter, we do not comment on the validity of these and other assumptions, which will be left to the concluding chapter. We shall only note, however, that this supply-side premise of deterrance is at variance with much of the prevailing deterrence theory, which is premised on mutual limitation of nuclear power (as opposed to arms race) and a credible second-strike capability (hence, deemphasis on first use), which also contains an appreciation for "overkill."[39]

Readiness to Use Strength

The second generalizable theme is that the United States must be ready to use its much augmented strength, as can be seen in the official statements already quoted. Although logically connected with this theme is the premise that nuclear war is winnable, the Reagan Administration officially denies this belief, imputing it instead to Soviet intentions. At his October 1, 1981, news conference, President Reagan answered a question in exactly that vein:

> It's very difficult for me to think that there is a winnable nuclear war, but where our great risk falls is that the Soviet Union has made it very plain among themselves that they believe it is winnable. And believing it, that makes them constitute a threat....[40]

The assumption that the Soviets believe in a winnable nuclear war may or may not be a projection of his own beliefs. It has, nevertheless, become a justification for U.S. preparation for a nuclear showdown, and by extension, a possible first use of U.S. nuclear missiles in a preemptive war against the Soviets. Responding to the Reagan charge, however, Leonid I. Brezhnev, the late Soviet leader, disclaimed any aggressive (first-use) designs. "Our military doctrine has a defensive character," he declared, adding: "It excludes preventive wars and the concept of a first strike."[41]

There probably was an initial confusion within the Administration between "first use" and "first-strike capability." In the wake of the celebrated *Foreign Affairs* article by McGeorge Bundy, George F. Kennan, Robert S. McNamara, and Gerard Smith, arguing for U.S. adoption of a "no first use," Administration officials backed into an awkward position of having to defend the first-use tenet.[42] For example, General Bernard

W. Rogers, Supreme Allied Commander/Europe, was obliged to refute the value of a declared "no first use" policy. "The single most important factor," he argued, "in restraining Soviet aggression will always be that chasm of uncertainty about Western readiness to cross the nuclear threshold."[43]

The readiness to use nuclear might in a possible first strike carries with it a necessary modification of the long-standing premise of deterrence. Instead of the prerequisite of MAD (mutual assured destruction), where the emphasis is on a credible second-strike capability, supply side deterrence involving possible first use requires a self-assuring augmentation of U.S. strategic forces (MX, B-1, cruise missile, etc.), a reliable targetting accuracy and silo-penetrating power of our warheads, and a foolproof protection of the land-based missile force (e.g., an ABM system). After all, according to this view, the Soviets will be made to behave only when they know we have overwhelming strength and we are ready to use it. Instead of basing deterrence on MAD, the Administration stakes out a strategy that embraces what some scholars call the "NUTS (nuclear utilization theories) logic."[44]

Multiple-Front Scenario

U.S. strategic thinking since World War II, until 1972, focused primarily on defense of Europe and Northeast Asia, involving a scenario of "2½ wars," that is, coping with the Soviet threat in Europe and a Sino–Soviet threat in the Far East, plus a limited war somewhere in the Third World. Exploiting the Sino–Soviet split and a growing U.S.–Chinese rapprochement, the Nixon–Ford Administration was able to accommodate to the post-Vietnam defense cutbacks by sizing U.S. forces to only "1½ wars."[45] The crises in Southwest Asia, in 1979–1980, which necessitated the Carter Doctrine and the inauguration of the RDF in their wake, pointed up the sad inadequacy of the "1½ war" scenario and the force levels supporting it. The Reagan team is pushing for a multiple-front scenario, going even beyond the erstwhile notion of "2½ wars."

"We may be forced," stated Secretary Weinberger, "to cope with Soviet aggression, or Soviet-backed aggression, on *several fronts*." "But," he added, "even if the enemy attacked at only one place,...we might decide to stretch our capabilities, to engage the enemy in *many places*, or to concentrate our forces and military assets in a few of the most critical arenas."[46] The preparation for "several fronts" is known as the doctrine of "simultaneity," and the readiness to take on the enemy not just at the theater of his choice, but in many places at the same time, is called "horizontal escalation." Eastern Europe may be the Soviet Union's Achilles heel. Cuba could also be a target against which U.S. military preponderance could be used in an instance of horizontal escalation.[47]

Senior Reagan officials conceive of "simultaneity" as a basic distinction between their strategy and that of the Carter Administration, which, they said, was willing to accept a smaller U.S. force, partly on the assumption that U.S. forces could be shifted from one theater to another, dealing with threats one at a time. What is the difference between simultaneity and horizontal escalation? The latter is the result of the U.S.'s own initiative, whereas the former may be forced by Moscow's choice of starting a war at several points simultaneously. The Reagan people also stress the importance of the ability of U.S. forces to carry out combat operations for months, rather than weeks, under the doctrine of "sustainability."[48] All these doctrines are calculated to forewarn the Soviets not to venture a war with the United States, which is prepared for all contingencies. "A wartime strategy that confronts the enemy, were he to attack, with the risk of our counteroffensive against his vulnerable points strengthens *deterrence*," explained Secretary Weinberger.[49] Thus, supply side deterrence has its application in conventional warfare as well.

A Neo-Mahan Strategy (?)

Although not yet consciously formulated as such, the maritime emphasis in the Reagan defense program seems to point to a strategy of swarming the Eurasian landmass (Heartland) power of the Soviets and confronting their recent maritime thrusts on and in the oceans. Alfred Thayer Mahan, the nineteenth-century naval officer *cum* historian, had advanced the thesis of the strategic advantages of sea power, which put the spotlight on naval strength, bases, control of sea routes, sea defense of metropolitan centers, and overseas sources of raw material supply.[50] Subsequent exponents of this thesis have also added that the global mobility of its sea power gives the West advantages of strategic initiative over a relatively immobile Heartland power (as defined by Mackinder). But, until the Soviets have established themselves as a global sea power, as well as land power, U.S. strategy has been dominated by the rimland concept, with its focus primarily on Western Europe and Northeast Asia.

The neo-Mahan addendum, which combines the strategic importance of the rimlands and of the oceans, does not suggest an abandonment of the Rimland strategy, only a modification. It is compelled both by the insufficiency of the West's previous concerns of containing the Heartland threat merely from the fringes of the "world island" (Mackinder's term) and by the breaking out of Soviet power through its strategic naval arm. According to Secretary of the Navy John F. Lehman, Jr., the rapid growth of Soviet naval power has "eliminated the option of planning for a regionally limited naval war with the Soviet Union."[51]

ASIAN POLICY IN THE GLOBAL STRATEGIC CONTEXT

Management of U.S. security interests in Asia under the Reagan Administration, the topic for the present volume, must therefore be examined within the larger global strategic picture we have just sketched. To reiterate, the Asia–Pacific and the Persian Gulf regions, both in larger Asia, constitute two of the three main fronts where the United States is endeavoring to constrain the Soviet threat, the third being Western Europe. As we have seen, larger Asia is important to U.S. strategic interests for three reasons: traditional U.S. concerns, vital resources (including oil), and the expansion of Soviet military might. The neo-Mahan strategy underscores the increasing strategic salience of the Indian and Pacific oceans and the South China Sea. Reagan's concerns for protecting the sea lanes, both those linking the sea areas and those through them, have economic as well as military overtones.[52]

To date, Soviet Pacific forces are based primarily in the Soviet Far East territories, including the new and permanent bases on some of the disputed Kurile Islands near Japan's northernmost Hokkaido Island, not to mention the SLBM sanctuary of the Okhotsk Sea. Most of these forces, plus the 51 army divisions along the Sino–Soviet border, are probably, for the moment, more directed toward China than any other country, although from the Okhotsk Sea Soviet submarines can launch ballistic missiles against the continental United States with relative impunity. Despite the Chinese manpower advantage, the Soviet superiority vis-à-vis the Chinese will probably not be altered in the foreseeable future, even with the levels of support the United States is willing and able to provide (see Robert G. Sutter's chapter for a discussion of U.S. debates on this topic). Questions surrounding the durability of the Sino–Soviet split raise further questions on the extent to which the United States can fully and indefinitely use the "China card" to its advantage (Carol Lee Hamrin's chapter addresses the possibility of a Chinese shift in their relations with the two superpowers).

Korea is also vulnerable to possible Soviet intervention on behalf of Pyongyang in the event of a military confrontation with South Korea. The Soviet Pacific Fleet's home port, nearby Vladivostok, affords Moscow a decided advantage in any such eventuality, despite its trans-Siberian logistical constraints. Hence, South Korea is an important strategic link for Reagan's game plan, not just an outpost for Japan's defense, as under Carter. In the event of a regional flare-up, Soviet naval craft based in Vladivostok have to pass through one of the three narrow straits (Tsushima, Tsugaru, or Soya), which can be blocked off by Japanese defense forces. Japan's role and commitment in this respect would be decisive for the outcome. In addition to its economic importance, Japan therefore has an additional strategic value that is only magnified in the eyes of a U.S. President engrossed with the Soviet problem (Norman Levin's chapter deals with these aspects of Northeast Asia).

The more recent expansion of Soviet military presence in Indochina, following the signing of the Soviet–Vietnamese friendship treaty in 1978, is one additional item on the U.S.' problem list for Asia. The Soviets, who have always been searching for more suitable bases and warm-water ports with access to the open sea, are now enjoying their use of former U.S. bases of Danang and Cam Ranh Bay. Though it is generally assumed that the Soviets are reluctant to be directly involved in areas remote from Soviet borders, the potential role they could play in Indochina and surrounding areas, due to their access to the Vietnamese bases and through them to the South China Sea and the western Pacific, cannot but be a source of concern for Washington, especially in view of the Soviet behavior in the Sino–Vietnamese confrontation. Soviet intervention in that instance suggests a readiness on Moscow's part to commit its military resources thousands of miles away from Soviet territories when a regional power play called for it, although the experience also reveals the limitations of its military posture in the Far East (John Garver's chapter). Similarly, continuing Soviet occupation of Afghanistan and their proximity to the Persian Gulf and redoubled Soviet aid to Syria since the latter part of 1982, cannot but sustain Washington's uneasiness and keep alive the Carter Doctrine without Carter (Winberg Chai's chapter thus explores the new prospects, following the Lebanese invasion, of a continuing U.S. Middle East peace plan, which conceives of Saudi Arabia as the "core" of Persian Gulf stability).

Severe logistical problems and a generally unfavorable Asian environment may have thus far prevented the Soviets from translating their enormous military power into political influence in most parts of Asia outside Indochina.[53] Nevertheless, for the Reagan Administration, the Soviet preoccupation has both elevated the strategic importance of Asia and colored its policy in return, in that the military/security reflex seems to shape or dominate all political considerations.

NOTES

1. Leigh S. Ratiner, "The Law of the Sea: A Crossroads for American Foreign Policy," *Foreign Affairs*, Vol. 60, No. 5 (Summer, 1982), pp. 1006–1021.

2. President Reagan paid the interest due on Polish debts shortly after the martial law was declared and was in full effect in Poland. See a discussion on this point in Norman Podhoretz, "The Neo-Conservative Anguish over Reagan's Foreign Policy," *New York Times Magazine*, May 2, 1982.

3. For a good discussion on the management of U.S.–Soviet relations, see William Hyland, "Avoiding a Showdown," *Foreign Policy*, No. 49 (Winter, 1982–1983), pp. 3–19.
4. Also Secretary of Defense Caspar Weinberger's *Report to the Congress for Fiscal Year 1983*.
5. President Reagan's news conference of March 31, 1982, *New York Times*, April 1, 1982.
6. Interview in *Wall Street Journal*, May 6, 1980.
7. Congressional Quarterly, *Weekly Report*, April 3, 1982, p. 767. Also, for a similar view expressed by General John W. Vessey, the new Chairman of the Joint Chiefs of Staff, see ibid., May 15, 1982, p. 1140.
8. Congressional Quarterly, *Weekly Report*, April 24, 1982, p. 947.
9. Secretary Weinberger's report, pp. 1–17.
10. *New York Times*, November 23, 1982, p. 1; *U.S. News and World Report*, December 6, 1982, p. 22.
11. Congressional Quarterly, *Weekly Report*, February 13, 1982, p. 248; idem, *1981 Almanac*, p. 195; *U.S. News & World Report*, January 10, 1983, p. 17.
12. Charles A. Monfort, "An MX 'Dense Pack' Would Need ABM's, Both Periling Security," *New York Times*, Op. Ed. page, December 1, 1982.
13. Congressional Quarterly, *Weekly Report*, February 13, 1982, pp. 248f.
14. U.S. Department of State, Bureau of Public Affairs, "START Proposal," July 1982; Richard Burt, "The Evolution of the U.S. START Approach," *NATO Review*, Vol. 30, No. 4 (1982), pp. 1–7.
15. Burt, *ibid.*, p. 5.
16. *Ibid.*
17. President Reagan's address before the National Press Club, Nov. 18, 1981, in New York *Times*, November 19, 1981.
18. Secretary of State Haig's address at the Georgetown University Center for Strategic and International Studies, April 6, 1982, *New York Times*, April 7, 1982.
19. U.S. Department of State, Bureau of Public Affairs, "Arms Control: MBFR Talks," October, 1982.
20. Iklé's appearance before a Senate subcommittee, on February 26, 1982, Congressional Quarterly, *Weekly Report*, April 10, 1982, pp. 795f; also Secretary Weinberger's annual report for fiscal year 1983.
21. Nicholas Spykman, *Geography of the Peace* (New York: Harcourt and Brace, 1944), pp. 37–41.
22. Cf. James C. Hsiung and Winberg Chai (eds.), *Asia and U.S. Foreign Policy* (New York: Praeger, 1981), pp. 191–208; 231ff; Colin S. Gray, *The Geopolitics of the Nuclear Era* (New York: Crane, Russak, 1977), p. 25.
23. Halford Mackinder, *Democratic Ideals and Reality* (New York: Norton, 1962), p. 277.
24. "Reagan Defense Plan Stresses Deterring the 'Soviet Threat'," Congressional Quarterly, *Weekly Report*, April 10, 1982, p. 792.
25. *Ibid.*
26. Peter Berton, in Hsiung and Chai (eds.), *Asia and U.S. Foreign Policy*, p. 12.
27. John Edwards, "Washington's Pacific Thrust," *Far Eastern Economic Review*, June 13, 1980, p. 40; also Dept. of Defense *Annual Report FY1982*, p. 173.
28. Berton, *ibid.*
29. *Jane's Fighting Ships, 1982–1983*, p. 407; and IISS, *The Military Balance, 1982–1983* (London, 1982), pp. 10, 13.
30. Donald S. Zagoria (ed.), *Soviet Policy in East Asia* (New Haven, Conn.: Yale Univ. Press, 1982), pp. 18, 270f.
31. Carl G. Jacobsen, "Soviet-American Policy: New Strategic Uncertainties," *Current History*, October 1982, p. 305ff.
32. Secretary Weinberger's report for FY 1983.

33. Reagan's address before the OAS, February 24, 1982, in Dept. of State, *Current Policy*, No. 370 (February 1982).
34. *New York Times*, October 30, 1980. Emphasis added.
35. Secretary Weinberger's annual report for FY 1983.
36. Secretary Haig, "A Strategic Approach to American Foreign Policy," Dept. of State, *Current Policy*, No. 305 (August 11, 1981). Emphasis added.
37. Secretary Shultz, "U.S. Foreign Policy: Realism and Progress, "Dept. of State, *Current Policy*, No. 420 (Sept. 30, 1982).
38. Address by William Clark before the City Club and Chamber of Commerce, San Diego, Calif., text in Dept. of State, *Current Policy*, No. 430 (Oct. 29, 1982).
39. Cf. William Epstein, *The Last Chance: Nuclear Proliferation and Arms Control* (New York: Free Press, 1976), pp. 28, 33, 201, 273.
40. *New York Times*, October 2, 1981.
41. *New York Times*, November 4, 1981.
42. McGeorge Bundy *et al.*, "Nuclear Weapons and the Atlantic Alliance," *Foreign Affairs*, Vol. 60, No. 4 (Spring 1982), pp. 753–768.
43. Bernard W. Rogers, "The Atlantic Alliance," *Foreign Affairs*, Vol. 60, No. 5 (Summer 1982), p. 1154.
44. Carl G. Jacobsen, *Current History*, note 31, p. 337.
45. Robert W. Komer, "Maritime Strategy vs. Coalition Defense," *Foreign Affairs*, Vol. 60, No. 5 (Summer 1982), p. 1127; Frank Church, "America's New Foreign Policy," *New York Times Magazine*, August 23, 1981.
46. Secretary Weinberger's report for FY1983. Emphasis added.
47. Congressional Quarterly, *Weekly Report*, April 10, 1982, p. 794.
48. *Ibid.*, pp. 794, 796.
49. Secretary Weinberger's annual report for FY1983. Emphasis added.
50. Alfred Thayer Mahan, *The Influence of Sea Power on History* (London: Sampson Low, Marton & Co., 1892).
51. *New York Times*, April 11, 1982, p. 1.
52. Zagoria, *Soviet Policy*, note 30, p. 27.
53. Cf. Paul Langer's chapter, in Zagoria, *Soviet Policy*, pp. 272, 277; also Zagoria's own chapter, esp. p. 4ff.

2

In Search of a Strategy: The Reagan Administration and Security in Northeast Asia

Norman D. Levin

INTRODUCTION

In listening to commentators on U.S. foreign policy under the Reagan Administration, it is hard to avoid the impression of a huge gap or polarization of opinion. On one side of the spectrum are Reagan critics who assail the Administration for lacking a coherent – or indeed *any* – foreign policy.[1] One the other side are Reagan loyalists who not only bristle at the suggestion that the Administration lacks a policy but aggressively assert that its selected policy is a major reversal of, or departure from, that of the preceding Administration. Indeed, spokesmen for the Administration have gone so far as to suggest that its efforts represent a coherent strategy designed to restore U.S. leadership and chart a new course for the United States in the 1980s.[2]

Whatever the situation elsewhere, on issues pertaining to security in Northeast Asia, neither contention is very convincing, nor very useful as a description of the management of U.S. interests in Asia under the Reagan Administration. On the one hand, the Reagan Administration clearly has a broad policy it is pursuing in Northeast Asia – if by "policy" one means a definition of national *interests* or *objectives*, an explication of the *requirements* needed to achieve these objectives, and the adoption of consistent *measures* to satisfy the stated requirements. On the other hand, the resulting "policy" of the Reagan Administration does not represent a dramatic reversal of, or departure from, that of the previous Administration – least of all a new "strategy" for the United States in the

---

The views expressed in this chapter are those of the author and do not represent the views necessarily of either RAND Corporation or its sponsors.

1980s. Rather, it represents an evolutionary extension of policy trends evident since at least the last 2 years of the Carter Administration.

The purpose of this chapter is to provide an alternative framework for viewing U.S. Asian policy. To provide such a framework, an attempt is made to identify the basic objectives the Reagan Administration is pursuing in Northeast Asia, to analyze the Administration's definition of what is required to achieve these objectives, and to describe the measures adopted by the Administration to satisfy the requirements defined. In this way, the intent is to provide a better basis for assessing the management of U.S. security interests in Asia under the Reagan Administration.

REAGAN AND NORTHEAST ASIA— THE EVOLVING SECURITY POLICY

Basic Objectives

The basic objectives being pursued by the Reagan Administration in Northeast Asia are both interrelated and a reflection of the Administration's larger global and regional concerns. Broadly speaking, these objectives are threefold in nature. The first is *to end what the Administration considers a decade of ambivalence and vacillation in Asia.* Criticizing the "zigzags," "inconsistency," and general "undependability" of previous administrations, the Reagan Administration has placed "consistency" and the need to demonstrate "loyalty" and "commitment" to U.S. allies and friends at the top of its policy priorities.

The second basic objective is *to check Soviet expansion.* Assailing the policy of "detente" and the notion that there is a natural community of shared interests between the United States and the Soviet Union, the Administration has adopted a more openly confrontational approach in dealing with the Soviets. In the process, it has made clear that checking Soviet expansion will be—in Asia as elsewhere—the paramount U.S. objective.

The third basic objective of the Reagan Administration is *to reassert American leadership in the region.* Through revitalizing relations with key U.S. allies such as South Korea and Japan and building expanded relations with nations like China, the Administration seeks to substitute for the previous U.S. "retreat" and general policy retrenchment a new, more activist role throughout East Asia. This role is both more assertively predicated on perceived U.S. self-interest and in keeping with the President's apparently deeply held convictions concerning America's proper place internationally as the "leader of the Free World."[3]

Central Requirements

To achieve these basic objectives, the Administration has identified a number of central requirements. First is a renewed and unequivocal American commitment to U.S. treaty allies, and a willingness to help victims or possible targets of "aggression" by the Soviet Union or its proxies. This is based on the perceived need to end previous American ambivalence and vacillations, and to demonstrate loyalty and commitment to U.S. friends and allies.

Second is a dismantling of previous restraints on overseas arms sales, and an expansion of both U.S. military aid and security-related economic assistance. Aiming at the enhancement of the state of preparedness of U.S. friends and the revitalization of U.S. alliances, this is predicated on the belief that "the strengthening of other nations with which we share common security interests is an essential component of our total effort to restore effective deterrence to aggression."[4]

Third is a substantial increase in U.S. military efforts, and a greater military emphasis in U.S. policy deliberations. This involves not merely increased military spending but also an expanded military doctrine that envisions protracted war with the Soviet Union extending to many parts of the globe at the same time as the major danger for which the United States must prepare.[5] It also involves heightened influence for the armed forces in plotting U.S. strategies, as well as an effort to "reinvigorate" the intelligence community by loosening past restrictions.

Fourth is the building of a loose grouping or coalition of friendly powers in the region. This involves the forging of a "strategic consensus" designed to broaden economic, political, and military cooperation against Soviet expansion.

Finally is a significant expansion of the U.S. posture in the Persian Gulf and Southwest Asia, and a more equitable "division of labor" and sharing of the defense burden between the United States and its principal allies. The former includes a major naval buildup and increased naval presence in the Persian Gulf area, the creation of an independent command for the RDF and plans to station combat troops in Southwest Asia, and broadened American participation in joint military exercises with a number of countries in the region. While clearly serving strictly *American* objectives, this expanded posture is designed to demonstrate U.S. recognition that the area is vital to *allied* interests, and that these allied interests will be defended, if necessary, by the United States.[6] The latter policy requirement involves a recognition of the constraints on U.S. resources, and the need for U.S. allies and friends to "join us in contributing more to the common defense."[7] Both reflect the perceived need to reassert U.S. leadership and to assume a more activist regional role.

Policy Measures

In its first 18 months, the Reagan Administration has adopted a number of specific measures to satisfy these requirements and achieve the basic objectives. This is immediately evident in the case of South Korea. As suggested succinctly in the Republican Party Platform, the policy of the Administration has been clear from the very beginning:

> Republicans recognize the unique danger presented to our ally South Korea. We will recognize the special problems brought on by the subversion and political aggression from the North. We will maintain ground and air forces in South Korea, and will not reduce our presence further. Our treaty commitments to South Korea will be restated in unequivocal terms and we will reestablish the process of close consultations between our two governments.[8]

From this position, President Reagan invited President Chun to Washington at the very beginning of his administration. On this occasion, he promised to maintain, and perhaps even to augment, the strength of U.S. forces in South Korea, and to give U.S. Pacific commitments the same weight as its European alliances. He also promised that the Unites States would rule out any bilateral discussions with North Korea unless the South were a "full participant."[9]

Indicating that the U.S. recognized the strategic value of the Korean Peninsula for its own sake rather than merely in terms of the defense of Japan, the Reagan Administration then proceeded to take a number of further measures. It decided not only to keep U.S. forces in South Korea but to strengthen these further by improving artillery and helicopter equipment, replacing older war planes with F-16s, and deploying a squadron of the most advanced U.S. ground support plane, the A-10, with American forces in South Korea.[10] In addition, the Administration promised to substantially improve U.S. military capabilities and assets in the region as a whole through such means as the deployment of the battleship *New Jersey* to the Western Pacific.[11]

To support South Korea's indigenous military modernization programs, the Reagan Administration also agreed to provide expanded military assistance—including appropriate sophisticated technology, advanced equipment sales, and improved foreign military sales (FMS) credits ($210 million in fiscal year 1983, an increase of $44 million or 26%). In line with this the Administration agreed to: sell South Korea at least 36 F-16s; provide South Korea the Stinger, the most advanced shoulder-launched antiaircraft missile; turn over to South Korean forces a Hawk antiaircraft system previously operated by U.S. forces; deliver F-5 jet fighter "parts" for aircraft manufacturing in South Korea; and transfer to South Korea a 4,500-ton destroyer. Beyond these measures, the U.S. signed an agreement at the annual U.S.–South Korea security consultative meeting in March 1982 that provides for the expeditious transfer to South Korea in an emergency of about $2 billion worth of war reserve stockpiles owned by the United States but stored in South Korea. This unique agreement was designed both to increase the immediate combat readiness of Seoul and to send a signal to Pyongyang.[12]

Added to these measures have been a range of security-related economic concessions. The United States agreed to postpone by 3 years, for example, the start of repayment on about $300 million in loans to buy arms sold to Korea in fiscal years 1981 and 1982, and pledged to ask Congress to permit similar concessions in subsequent years. The Administration gave tentative approval for South Korea to sell six kinds of military equipment made in that country under U.S. license to 24 smaller third world nations. The Administration also offered to sell South Korea over 1,000 M-55 light tanks for $10,000 each, a fraction of their original cost.[13]

Finally, the Reagan Administration reaffirmed the nuclear umbrella for South Korea. In what was perhaps the strongest expression of the U.S. security commitment to South Korea in a dozen years, the United States labeled the security of Korea "pivotal" and pledged to render "prompt and effective assistance to repel aggression against the Republic of Korea."[14] Adoption of these measures toward South Korea was designed to show an unequivocal commitment to U.S. treaty allies and to demonstrate that friends of the United States can expect to receive a sympathetic hearing in Washington. As such, they accurately reflected the Reagan Administration's policy toward Asia in general and toward Northeast Asia in particular.

The measures adopted toward Japan have been similar in nature but designed to address the two other perceived policy requirements, namely: to further the building of a loose grouping or coalition of friendly powers to contain Soviet expansion; and to provide a more equitable "division of labor" and sharing of defense burden. Between the two the latter has been more aggressively sought. From the beginning, the Reagan Administration made it clear that it considers Japan a "full partner" of the United States, and that a "full partnership" involves both rights and responsibilities. This perception has heavily colored the Administration's approach. Rather than pressure Japan on specific issues such as the percentage of the budget or GNP allocated to defense, the Administration has sought to engage Japan in a broad discussion of security needs in the 1980s. The purpose of such a discussion has been to establish *common goals* concerning defense, trade, and energy, and the establishment of world order.

On this basis, the Administration has hoped that Japan would willingly assume greater responsibility for the achievement of these common goals.

In his first major policy speech, billed as an exposition of U.S. policy for East Asia, Secretary of Defense Weinberger laid out the Administration's twin themes. On the one hand, the United States would remain a loyal and reliable ally. Recognizing that the U.S. military presence in Asia is indispensable to Japanese security, Weinberger pledged that the United States "will continue to provide the nuclear umbrella and offensive striking power in Japan's behalf."[15] On the other hand, however, the increasing threat "clearly requires significantly greater self-defense efforts in the Northwest Pacific."[16] Focusing on Japan's share of the burden in making these greater efforts, he explained that a "cornerstone" of U.S. policy would be "to develop a rational division of some of these burdens so that our allies, with their own enormous strengths, can join together with us in contributing more to the common defense."[17]

On the part of the Reagan Administration, the summit meeting between the President and Prime Minister Suzuki in Washington in early May, 1981 represented an effort to "align the perspectives" of the two countries on "the growing Soviet strategic threat" through broad and general discussions. The joint communique issued after these discussions indicated substantial success in this effort. In the joint communique Suzuki agreed: that ". . . the alliance between the United States and Japan is built upon their shared values of democracy and liberty;" that the "determined efforts" of the United States in the Mideast "contribute to restoring stability and that many countries, including Japan, are benefitting from them;" and that in meeting the "international challenges to their peace and security, all Western industrialized democracies need to make greater efforts in the areas of defense, world economic improvement, economic cooperation with the third world, and mutually supportive diplomatic initiatives."[18] Suzuki also acknowledged "the desirability of an appropriate division of roles between Japan and the United States," and pledged that Japan ". . . will seek to make even greater efforts for improving its defense capabilities in Japanese territories and in its surrounding sea and airspace, and for further alleviating the financial burden of U.S. forces in Japan."[19]

Before leaving Washington, Prime Minister Suzuki publicly stated Japan's intention to "defend its own territory, the seas and skies around Japan, and its sea lanes to a distance of 1,000 miles" in a speech before the National Press Club.[20] Using this statement, together with the joint communique, as a lever, the Reagan Administration has sought to persuade the Japanese to expand their defense responsibilities and to speed up their defense buildup. With the objective of minimizing the Soviet naval threat to American sea and air lines of communication and retarding the expansion of Soviet influence in the region, the Administration has sought to get Japanese agreement to three specific roles and missions in particular: patrolling the seas and skies west of Guam and north of the Philippines in order to keep the sea lanes open; acquiring ships, aircraft, and mines able to close off the straits leading from the Sea of Japan to the Pacific Ocean in order to bottle up the Soviet submarine and surface fleet at Vladivostok; and stockpiling ammunition, supplies, and other logistical support for the ships, planes, and other weapons currently in the Japanese armed forces.[21] Japan's decisions to increase defense spending in fiscal year 1982 by 7.75% and to approve a 5-year defense program estimated to cost $62–$65 billion between 1983 and 1987 – including an $18 billion increase in spending on advanced weapons – represent initial steps in developing such a capability and fulfilling the objectives laid out by Prime Minister Suzuki.[22]*

At the same time, the Administration has urged Japan to increase its economic aid to both South Korea and China in ways that would allow those nations to use their money for greater military spending. Indeed, the Administration has described increased Japanese foreign assistance as one of the main "policy objectives" of the United States. Assistant Secretary of State for East Asian Affairs John Holdridge, for example, in a statement before the Subcommittee on Asian and Pacific Affairs of the House Committee on Foreign Affairs on March 1, 1982, made note of the fact that Japan has been "increasingly willing to provide significant amounts of aid, often fast-disbursing to countries of political importance to the Western alliance, even if they are of relatively little economic importance to Japan." The U.S. believes, Mr. Holdridge added approvingly, that "prospective future aid developments in Japan will satisfy both Japan's own interest and support U.S. objectives."[23]

All this is in line with the perceived need to have U.S. allies contribute more to the common defense, and to build a loose coalition of friendly powers to contain further Soviet expansion in Asia. While the U.S. commitment to help defend Asia is solid, the Administration has emphasized, the region is so vast that more extensive Japanese efforts and more effective Japanese forces are needed for the common defense and the overall stability of the region.

---

* Editor's note: The situation has not changed greatly since Prime Minister Yasuhiro Nakasone visited Washington in early 1983 (see Chapter 8).

POLICY AND STRATEGY–THE MISSING LINK

The Policy and the Past – Change or Continuity

If this analysis is roughly accurate, then it seems fair to acknowledge that the Administration is indeed pursuing a broad "policy" in the security area in Northeast Asia. Care must be taken, however, not to exaggerate the "novelty" or "newness" of the policy selected. One obvious example concerns military spending. Although the Administration has embarked on a substantial expansion of U.S. military capabilities, the trend in this direction had already been clearly set by the Carter Administration. In President Carter's last budget, for example, he proposed real increases in defense spending of 7.8% for fiscal year 1981 and 5.3% for 1982. The fiscal year budgets proposed by the Reagan Administration for its first 2 years are at best modest increases over the trend initiated under President Carter.[24]

A similar situation exists concerning the Reagan Administration's emphasis on checking Soviet expansion. This rhetorical and even doctrinal emphasis has given the Administration something of a "confrontationist" image, and subjected it to substantial criticism both at home and abroad. Despite this clear emphasis, however, it is not in fact the Reagan Administration that rediscovered "containment" but the administration of President Carter. As early as the spring of 1978, for example, President Carter warned that the Soviet Union was defining detente as "a continuing aggressive struggle for political advantage and increased influence."[25] "The Soviet Union apparently sees military power and military assistance as the best means of expanding their influence abroad," the President continued, and with "areas of instability" providing "a tempting target," the Soviets "all too often... seem ready to exploit any such opportunity."[26]

From this perspective, President Carter reversed not only the steady decline in real defense spending, as suggested above, but also his previous antipathy to arms sales which he proceeded to significantly increase.[27] In what one observer recently called the "most farreaching extension of American commitments since the redefinition of America's Pacific defensive perimeter after the Korean war," President Carter formally extended U.S. defense responsibilities to Southwest Asia following the fall of the Shah of Iran and the Soviet invasion of Afghanistan.[28] Subsequently dubbed the "Carter Doctrine," the United States pledged that "an attempt by any outside force to gain control of the Persian Gulf region will be regarded as an assault on the vital interests of the United States of America, and such an assault will be repelled by any means necessary, including military force."[29] Going beyond mere rhetoric, the Carter Administration established a "Rapid Deployment Force," deployed naval forces to protect North Yemen from its rival in the South, imposed a partial grain embargo and other sanctions on the Soviet Union, and moved to significantly bolster relations with the People's Republic of China, including the removal of some restrictions on sales of militarily significant technology, as a means for countering Soviet expansion.[30] The concomitant lowering of the relative priority of East Asia in U.S. military planning, and the Administration's parallel efforts to persuade U.S. allies, particularly Japan, to increase their own defense efforts, represented a final legacy bequeathed to the Reagan forces.

None of this, it should be emphasized, is meant to suggest that there are no important differences between the two Administrations. Rather it is simply meant to suggest that there is more continuity to U.S. policy than may at first meet the eye. In analyzing the security policy of the Reagan Administration in Northeast Asia, this seems an important message to keep in mind.

The Policy and the Present—The Missing Link

Although it is still too early to assess fairly the efficacy, let alone wisdom, of the Reagan Administration's policy in Northeast Asia, it appears already to have had a number of positive effects. Chief among these has been a termination of the ambivalence of much of the previous period concerning the nature of the U.S. commitment to South Korea, and a solidification of U.S.–South Korea ties. This has been conducive to lowering South Korean and Japanese anxieties and helping further the possibility for greater stability in Northeast Asia.

Despite such positive effects, however, the Administration has been plagued by an apparent inability to fashion and articulate an overall *strategy* for achieving its policy objectives. That is, while the Administration has generally succeeded in defining national objectives and identifying policy requirements, it has thus far failed to establish a comprehensive plan that indicates priorities among these objectives, and that integrates these priorities with available resources.

This failure to fashion and articulate a clear, overall strategy for achieving policy objectives is reflected in a number of policy dilemmas. Most obvious is the conflict between the goal of expanding U.S.–China political and military ties in an effort to build a loose coalition to contain Soviet expansion, on the one hand, and the desire to demonstrate steadfastness and loyalty to Taiwan, on the other. Despite the efforts of the Administration to finesse this issue—most notably in the personal letters from President Reagan addressed to top Chinese leaders—the fundamental conflict remains unresolved.[31]

More specifically in Northeast Asia, the need for a new commitment and for "rewarding" U.S. allies such as South Korea through a strengthened military presence and greatly expanded military assistance conflicts, at some level, with the desire to lower tensions on the Korean peninsula and prevent the further expansion of Soviet influence through bolstered Soviet ties with North Korea. Similarly, the desire to impress upon the Japanese the need for greater "burden sharing" risks a potential backlash that would precipitate precisely the opposite reaction: namely, a move by Japan away from the Western alliance and toward a more "independent" policy position. All of these suggest the need to develop a broader *strategy* for Northeast Asia that coherently explains what U.S. priorities are, how these priorities relate to other East Asian and global objectives, and the manner in which they can be integrated with both allied interests and available U.S. resources.

A related point concerns the question of the place of Asia in overall U.S. planning. As described in newspaper accounts of Pentagon "defense guidance" planning, Asia now formally ranks behind the continental United States, Western Europe, and the Persian Gulf area in defense planning priorities.[32] To be sure, some lowering of the relative priority of Asia is understandable, and in any case probably inevitable, given the constraints on U.S. resources and the pressing demands on the United States in vital and unstable parts of the world. Any plans for a major shift of U.S. forces from the western Pacific to other regions, however, pose potential problems at two principal levels. Politically, such plans could seriously undermine the confidence of key Asian allies such as Japan in the credibility of the U.S. defense commitment. Were this to happen, as suggested above, the effect would very likely be to strengthen neutralist sentiment in Japan and significantly diminish Japanese enthusiasm for any U.S.-sponsored anti-Soviet endeavors. Even in strictly military terms, however, such plans pose certain dilemmas for the United States given the threat of Soviet submarine-launched ballistic missiles (SLBMs) in any truly "global war" scenario. This threat suggests the need for a continuous military presence and capability in the northwestern Pacific region that appears to some degree to conflict with evolving Administration policies. Despite the obvious constraints on U.S. resources, therefore, and the well-founded desire for greater Japanese defense contributions, it is essential that the importance of Asia, in general, and Japan, in particular, be very carefully factored into U.S. global planning.

A final dilemma resulting from the absence of an overall U.S. strategy concerns the interrelationship between economics, politics, and national security. As described above, the Reagan Administration policy in Northeast Asia has tended to approach the question of "security" primarily in *military* terms. While to some extent this is understandable given ongoing trends in the military balance between the United States and the Soviet Union, a singular fixation on the military dimensions of "security" obscures the manifold ways in which issues of a more economic or political nature can intrude and fundamentally affect security objectives. One obvious example concerns trade and the apparent collision course the Administration increasingly seems to have embarked on with key U.S. allies. Another example concerns arms control, and the political policies the Administration has adopted toward negotiations with the Soviet Union. The danger here is that U.S. allies—in the case of Northeast Asia, Japan in particular—might come to see the United States as not only insensitive to their positions but threatening to their fundamental interests. Should this occur the fallout on U.S. "security" objectives in even the strictly "military" sense would be substantial. Such a possibility suggests the need for giving greater emphasis to diplomacy in U.S. national strategy, and for carefully integrating economic and political measures into broader security concerns.

The Policy and the Future—Some Issues for the 1980s

Beyond such dilemmas associated with the absence of an overall strategy for achieving policy objectives, the strong emphasis of the Reagan Administration policy on checking Soviet expansion leaves largely unaddressed a range of broad policy issues that are likely to be of importance in the 1980s. One of these is the issue of moderating superpower competition and ultimately coming to grips with the Soviet Union as an Asian power. Given the continuing buildup of Soviet military capability and the demonstrated willingness of the Soviet Union to take advantage of available opportunities, the United States has a clear interest in opposing further expansion of Soviet influence. At some point, however, the United States will have to come to terms with the Soviet Union's emerging role as a regional power. How to assimilate the Soviet Union into Asia and influence it to play a positive rather than a disruptive role will likely constitute an important issue of the 1980s.

A second issue will be adjusting the growing interest in collaborative regional security arrangements to the realities of serious internal and external constraints—not to mention differing definitions of national security and divergent perceptions of national threat—that limit the alliance potential of various Asian states. To be sure, a U.S. interest in regional coalitions and collaborative security arrangements is understandable given the dilemma of limited resources and increased demands. The disparate nature of the various states of Asia, however, and the intraregional suspicions and hostilities that constrain their cooperation, significantly limit or circumscribe their alliance potential. What kinds of formal or tacit security arrangements can be worked out given these limits, what kind of "division of labor" is both appropriate and possible given these constraints, are questions that will require increasing attention.

A third broad issue will be that of managing the potential conflict between economic interdependence and growing economic nationalism. As growth rates remain modest and problems of inflation, unemployment, and trade deficits mount, economic nationalism and a desire for greater autarky will almost certainly proliferate. How to reconcile this with the reality of economic interdependence, and prevent the resulting conflict from spilling over into security areas, will likely preoccupy policymakers throughout the decade.

In addition to such broad security issues, there are a number of more specific ones relevant to U.S. bilateral relations in Northeast Asia. One key issue will be managing higher levels of tension in U.S.–Japan relations. This will undoubtedly involve channeling an increasing desire for "independence" on the part of the Japanese into a healthy regional role. Another will be encouraging South Korea to bring its political development into greater balance with its remarkable economic growth, both as a means for minimizing domestic instability and as a basis for expanding U.S. public support for a continued American security role. Still another issue will be creating incentives for North Korea to decrease its extraordinarily high level of military effort, and to take steps to reduce the level of tension on the Korean peninsula.

To state these issues is not, of course, to solve them—only to indicate something of the range of security-related matters with which the Reagan Administration will increasingly have to be concerned. This suggests the need to go beyond the Administration's present policy inclinations and to fashion a broader, more all-embracing strategy for the problems of security in Northeast Asia. It is to this task that U.S. policymakers will increasingly have to turn.

NOTES

1. See, for only one example, Joseph C. Harsch, "U.S. Foreign Policy: Signals, Sanctions, and Indecision," *Christian Science Monitor*, June 25, 1982.

2. See Secretary of State Haig's statement before the Senate Armed Services Committee on July 30, 1981 and his address before the American Bar Association in New Orleans on August 11, 1981, both of which are reproduced in the *Department of State Bulletin*, September, 1981.

3. For an early portrayal of these convictions, see the *Washington Post*, July 17, 1980. To these three objectives might be added one more: namely, to strengthen the domestic

economy. While not discussed in this paper, the Administration's awareness that a sound economy is a major component of national security, and its preoccupation with efforts toward furthering this end, should probably also be considered an aspect of its security policies.

4. James L. Buckley, Under Secretary of State for Security Assistance, Science and Technology, in a speech to the Aerospace Industries Association, quoted in the *Washington Post*, May 22, 1981.

5. Since there are innumerable discussions available on the substantial increase in military spending by the Reagan Administration, this issue is not discussed in detail in this paper. For useful summaries, see: "Reagan's Arms Buildup" in *Newsweek*, June 8, 1981, pp. 28–46; Herschel Kanter, "The Reagan Defense Program in Early Outline" in *Strategic Review*, Summer 1981, pp. 27–38; and Kevin Lewis, "The Reagan Defense Budget: Prospects and Pressures," P-6721 (The Rand Corporation, December 1981). For critical appraisals, see: James Fallows, "The Great Defense Deception" in the *New York Review of Books*, May 28, 1981, pp. 15–19; Hedrick Smith, "How Many Billions for Defense" in the *New York Times Magazine*, November 1, 1981; and Richard Stubbing, "The Imaginary Defense Gap: We Already Outspend Them" in the *Washington Post*, February 15, 1982. On the broadened military doctrine, see Secretary of Defense Weinberger's address before the American Newspaper Publishers Association in May 1981, the text of which is in the *Department of State Bulletin*, July 1981, pp. 46–47 and his *Annual Report to the Congress, Fiscal Year 1983*, pp. I/9–I/17.

6. As Secretary of Defense Weinberger put it: "Our vital interests are involved in Southwest Asia, as are, of course, those of our allies and of the independent states of that region. We will confront by force, if necessary, any Soviet or Soviet-inspired military threat to those combined interests. We are determined to demonstrate once again to our allies the reliability and value of American friendship. . . ." Caspar Weinberger, "The Defense Policy of the United States," *Nato's Fifteen Nations*, June–July 1981, pp. 14–16. The Administration has repeatedly made it clear that it is undertaking such "enormous increases in our efforts" with the full awareness that the areas of Southwest Asia "are of vital concern" to key Asian allies such as Japan. See, for example, the *Washington Post*, April 29, 1981.

7. Secretary of Defense Weinberger, quoted in the *New York Times*, March 25, 1981. Also see the *Wall Street Journal*, June 22, 1981.

8. *Far Eastern Economic Review*, November 14, 1980, p. 12.

9. *Los Angeles Times*, February 3, 1981.

10. *Washington Post*, April 29, 1981 and the *Los Angeles Times*, April 30, 1981.

11. *New York Times*, March 30, 1982.

12. *Washington Post*, April 1, 1982.

13. *Los Angeles Times*, April 30, 1981. The M-55 is considered obsolete but can be adapted for use as personnel carriers.

14. On the occasion of Secretary of Defense Weinberger's visit to Seoul in March, 1982 for the annual U.S.–Republic of Korea security consultative meeting, a joint communiqué was issued in which Weinberger "confirmed that the United States' nuclear umbrella will continue to provide additional security to the Republic of Korea." It is unusual for a joint communiqué to specifically call attention to American nuclear protection. *Washington Post, op. cit., April 1, 1982*.

15. Secretary of Defense Weinberger's speech to the Commonwealth Club and World Affairs Council in San Francisco, *Washington Post*, April 29, 1981.

16. *Ibid.*

17. *New York Times*, April 29, 1981. Citing the nuclear and conventional military protection the U.S. provides to Japan, including safeguarding of the Persian Gulf region upon which Japan is so dependent, Weinberger emphasized that "as in every true partnership, along with the benefits to be derived from the association, both partners incur obligations." *Los Angeles Times*, April 29, 1981.

18. The text of the joint communiqué is in the *New York Times*, May 9, 1981.
19. *Ibid.*
20. *New York Times*, May 9, 1981.
21. *New York Times*, May 5 and 8, 1981.
22. For details of the five-year defense program, see the *Washington Post*, July 24, 1982.
23. For the text of Holdridge's statement, see the *Department of State Bulletin*, April 1982, pp. 52–60.
24. President Reagan's proposed defense outlays for fiscal years 1981 and 1982 represented 5.6% and 5.8% of the gross national product, respectively. This compared with 5.5 and 5.6%, respectively, under the budgets proposed by President Carter. Dramatic differences in proposed budgets would not become evident until fiscal year 1985, and these would hinge on the state of the economy. *Washington Post*, June 7, 1981. For a skeptical view regarding the prospects of actually meeting the targets set in the budgets proposed for the later years, see Lewis, *op. cit.*
25. This was in a speech at the United States Naval Academy in June, 1978 – *before* the Soviet invasion of Afghanistan. For a transcript of the President's speech, see the *New York Times*, June 8, 1978, p. 22.
26. *Ibid.* The Naval Academy speech, incidentally, followed by less than three months a major Presidential speech at Wake Forest in which President Carter warned of the growth of Soviet military power and Moscow's "ominous inclination" to use this power to intervene in local conflicts. The President labeled as "myth" the notion that the U.S. was "somehow pulling back from protecting its interests and its friends around the world," and stressed that the U.S. was moving to improve both conventional and nuclear military capabilities in order to respond to Soviet actions. For excerpts from the speech see the *New York Times*, March 18, 1978.
27. Excluding training, U.S. military assistance delivered increased from $108.7 million in fiscal year 1977 to $336.7 million in fiscal year 1980. During this period, U.S. assistance to East Asia and the Pacific climbed from $44.6 million to $131.4 million. *Defense and Economy – World Report and Survey, Government Business Worldwide Reports* (Washington, D.C., 1982), p. 4655.
28. Robert E. Osgood, "The Revitalization of Containment," *Foreign Affairs (America and the World 1981)*, Vol. 6, No. 3, p. 650.
29. *Ibid*, p. 650.
30. *Ibid*, p. 650.
31. See the *Washington Post*, May 10, 1982, for a report on the Reagan letters.
32. *New York Times*, May 30 and June 7, 1982.

3

U.S.–China Military Ties: Implications for the United States

Robert G. Sutter

The Sino–U.S. reconciliation begun by President Richard Nixon and developed by succeeding U.S. presidents has won wide bipartisan support as one of the most important strategic breakthroughs in U.S. foreign policy since the Cold War. In broad terms, each U.S. Administration, from Nixon's to Reagan's, has sought to use better relations with China as a means to position the United States favorably in the U.S.–Soviet–Chinese triangular relationship; to stabilize Asian affairs, secure a balance of forces in the region favorable to the United States and China, and do so without damaging a peaceful and prosperous future for Taiwan; to build beneficial economic, cultural and other bilateral ties; and to work more closely with China on issues of global importance such as world food supply, population control, and arms limitations.

The motive force behind U.S. policy has been the search for strategic advantage. During the period of U.S.–Soviet detente in the first half of the 1970s, the United States sought to use improved relations with China as a means to elicit positive Soviet foreign policy behavior regarding U.S. interests; and to use improved relations with Moscow as a means to elicit positive Chinese foreign policy behavior. American planners sought to gain a position in which the United States would develop good relations with both communist countries, having better relations with them than they had with one another.[1]

In the face of the expansion of Soviet military power and political influence in such third world areas as Angola, the Horn of Africa,

---

The views expressed in this chapter are solely those of the author and do not represent the opinion of the Congressional Research Service or any other part of the U.S. Government.

Afghanistan and Indochina, and the continued steady growth of Soviet strategic and conventional forces, American interest in detente with the Soviet Union declined in the late 1970s. American policy was also affected by a rising U.S. concern over American military preparedness to meet Soviet and other foreign challenges seen following the collapse of the U.S.-supported governments in Indochina, the capture and detention of the American hostages in Iran, and the acrimonious debate over American strategic preparedness during the U.S. Senate deliberations concerning the SALT II Treaty.

As a result, the Carter Administration, especially in its last 2 years, shifted away from the policy of "even-handedness" that had characterized the American approach to the Sino–Soviet powers in the past. Improved relations with China increasingly came to be seen as an important source of regional and global power and influence for the United States, useful to America and its allies in the developed world as a means to counter what came to be seen as the major strategic problem for the next decade— the containment of expanding Soviet military power and influence in world affairs.

Officials in the Reagan Administration seemed to agree with this basic position. In an effort to consolidate ties with China, Secretary of State Haig traveled to Peking in June 1981 and announced that the United States, for the first time, was now willing to consider the sale of U.S. weapons to the People's Republic of China.

American support for closer strategic cooperation with China, nonetheless, began to splinter as U.S. officials moved beyond the establishment of normal Sino–U.S. political and economic relations and began to express willingness to sell military equipment to China.[2] By 1980, Sino–American relations had reached a new stage as contending groups within and outside the U.S. Government began to debate actively in public and private whether or not the United States should take what was widely seen as the next step forward in developing Sino–U.S. cooperation: the sale of weapons and weapons-related technology to China. The issue of whether or not to proceed with such sales was not only of military significance but had broader political importance as well. It directly impacted on a central question in U.S. foreign policy: how far should the United States go in trying to move closer to China in order to improve the American international position against the Soviet Union. Opinion has ranged widely. Some have judged that the United States has already gone too far in military ties with China and should stop promptly. Others generally have been satisfied with the existing policy but firmly resist further development, at least for the foreseeable future. Still others have favored a gradual increase in such transfers to China as an important step leading to the formation of a Sino–U.S. mutual security arrangement against the Soviet Union.

Continued strong differences over the sensitive Taiwan issue between the United States and China served to put at least a temporary cap on the developing strategic relationship.[3] Peking brought the American debate in this respect to a halt when it made clear in late 1981 that China would not move ahead with increased military cooperation with the United States until the Reagan Administration clarified its position on Taiwan. Heavy Chinese pressure forced Americans to focus attention on meeting Peking's demand that the United States agree formally to a gradual cutoff in U.S. arms sales to Taiwan. If the United States sold weapons to Taiwan without first agreeing to a gradual cutoff, Peking asserted that it would downgrade U.S.–China relations.

Faced with strong Chinese demands, the Reagan administration announced in January 1982 that it had decided not to sell Taiwan jet fighters more advanced than the F-5E in production in Taiwan, but it reaffirmed its commitment to sell arms, reportedly including more F-5E fighters, to the island. The Administration did not specify the number of U.S. planes to be sold or the duration of the sales. The decision ended months of debate in the United States on whether or not the United States should agree to sell a more advanced fighter, designated the FX, to Taipei, and it was widely hailed by the Western press as a sign of sensible moderation in U.S.–China policy.

China remained dissatisfied with the U.S. stance. After several months of reportedly delicate negotiations with the Chinese, the Reagan Administration announced in April 1982 that it was going to sell a package of military spare parts to Taiwan, the first sale of military equipment to Taiwan specifically authorized by the Reagan government. The Chinese did not downgrade relations because they said that the package did not involve weapons, had been agreed to prior to Chinese demands for a gradual cutoff of all arms, and – in the Chinese view – the United States had agreed not to consider military sales to Taiwan while Sino–U.S. discussions on the issue continued.[4]

Vice President George Bush traveled to Peking for talks in May 1982 and made public three letters from President Reagan to Chinese leaders linking a gradual decrease in U.S. arms sales to Taiwan with progress in Chinese efforts to reunify peacefully with Taiwan. Concerned with what they saw as the administration's deference to Chinese pressure at the expense of Taiwan's interests in delaying approval of more fighter aircraft to Taiwan, conservatives in the U.S. Congress and within the Republican Party began exerting a strong public pressure on the Reagan Administration during June and July for policy decisions favorable to Taiwan. Coincidentally, Secretary Haig resigned and was replaced by George Shultz, who said during his confirmation hearings before the Senate Foreign Relations Committee that he supported the sale of fighter aircraft to Taiwan.[5]

The United States and China on August 17, 1982 issued a joint communiqué that established at least a temporary compromise over the arms sales question. As a result, 2 days later the Administration was able to announce, without prompting a major, hostile Chinese response, the proposed sale of 60 F-5E/F aircraft to be coproduced in Taiwan from 1983 to 1985.

In the August 17, 1982 communiqué, the Reagan Administration said that U.S. arms sales to Taiwan will not exceed "either in qualitative or in quantitative terms" the level of those supplied during the past 4 years; and that the United States intends gradually to reduce its sales of arms to Taiwan, leading to what the communiqué called a "final resolution." The United States also disavowed a policy of "two Chinas" or "one China, one Taiwan," and said that the United States Government "understands and appreciates" the Chinese policy of seeking a peaceful resolution of the Taiwan question as seen in recent Chinese proposals for reunification talks with Taiwan. The U.S. side strongly implied that its agreement to curb arms supplies to Taiwan was contingent on a continuation of Peking's peaceful approach to the island.

For its part, Peking strongly affirmed its "fundamental" peaceful policy toward Taiwan. It also allowed the communiqué to go forward without reference to a fixed date for a U.S. arms cutoff, as the Chinese had demanded in the past. At the same time, Peking did not downgrade relations when the Reagan Administration formally notified Congress on August 19 that the United States would extend its current coproduction arrangement in Taiwan for F-5E/F fighters for over 2 more years.

Both sides averred in the joint communiqué that they would take unspecified future measures to achieve the "final settlement" of the issue of U.S. arms sales to Taiwan "over a period of time."

Subsequently, Chinese comment showed continued strong sensitivity over the Taiwan issues, demanding strict U.S. adherence to the communiqué and criticizing U.S. interpretations of the accord. An August 17 *People's Daily* editorial, commenting on the accord, warned that Sino–U.S. relations will "face another crisis" like the recent impasse over arms sales if U.S. leaders continue to adhere to the Taiwan Relations Act, which Peking now called "the fundamental obstacle to the development of Sino–U.S. relations."

If Peking were to reduce its pressure on the question of Taiwan and allow for greater development in U.S. military cooperation with China, American debate concerning the wisdom of increased U.S. defense ties and assistance to China would likely once again come to the fore. Trends in the triangular relationship among the United States, Soviet Union, and China would appear to support the case for increased U.S.–Chinese military ties. Pressure for American sales of military equipment would likely build so long as U.S. relations with the Soviet Union remain more hostile than cooperative, U.S. relations with China develop along the road to greater friendship, and Sino–Soviet relations remain cool.

BACKGROUND[6]

Although debate over U.S.–China military ties has only recently received prominent public attention, the roots of this debate go back to the Nixon Administration, when the pros and cons of seeking U.S.–China security cooperation were considered prior to significant U.S. initiatives toward China. These initiatives included the Nixon Administration's decisions to sell China a sophisticated ground station designed to pick up and transmit television signals via satellite, and several Boeing 707 aircraft with attendant aeronautical technology; and the Ford Administration's approval of the sale to China of British Rolls Royce "Spey" aircraft engines and related technology for China's fighter aircraft program, and its approval of the sale to China of an American computer that had potential military applications. Evidence of a growing U.S.–China security relationship was also seen in frequent joint consultations of Chinese and American officials on global and regional military issues, arms control, and other security matters, and in repeated American claims that it would strongly oppose any Soviet effort to dominate or establish "hegemony" over China.

The Carter administration continued this primarily symbolic interchange during visits of Secretary Vance and Dr. Brzezinski to China in August 1977 and May 1978, respectively. The Administration took other steps that incrementally increased Sino–U.S. security ties. For example, not only did American leaders repeatedly voice support for Chinese security against Soviet "hegemony," but they began in 1978 to stress American backing for a "strong" as well as for a "secure" China.

U.S. warships called on Hong Kong, which Peking regards as part of China, and received Chinese communist officials for well-publicized visits. The United States also adopted a more liberal attitude toward the transfer of sophisticated technology to China, for instance allowing China to purchase special U.S. geological survey equipment and nuclear power plant technology. Of more importance, the United States announced in November 1978 a shift from its past policy of opposition to Western arms transfers to the Sino–Soviet powers. It indicated that the United States would no longer oppose the sale of military weapons to China by Western European countries.

The strategic importance of Sino–U.S. ties was underlined by Defense Secretary Harold Brown during a visit to China in January 1980 – 1 month after the Soviet Union invaded Afghanistan. Headlines reporting on that visit and subsequent high-level interchange between military leaders of China and the United States spoke boldly of an emerging Sino–American "alliance" focused against the Soviet Union. Indeed, the rapid decline in Soviet–U.S. relations at the end of 1979 and the increasingly strong perception in the United States of a growing Soviet military threat seemed to give impetus to U.S. interest in closer ties with China, including security ties. Chinese leaders, for their part, stressed repeatedly their desire to develop a "long-term, strategic" relationship with the United States, adding on occasion their strong interest in obtaining weapons and other military equipment from the United States.

Following Secretary Brown's visit to China, the United States announced that it was willing to consider sales to China of selected military items and technologies with military support applications. In March 1980, the Administration listed categories of military support equipment the United States would consider, on a case-by-case basis, for export to China (Munitions Control Newsletter No. 81). Included on the list were radar, communications, and training equipment, trucks, transportation aircraft, and unarmed helicopters. The State Department announced that it would consider each export license application individually, bearing in mind the level of technology involved and the items' intended use.

Secretary Brown announced in mid-1980 that the U.S. Government had approved requests from several American firms to make sales presentations to the Chinese for certain articles of military support equipment and dual-use technology (items primarily of civilian use but with possible military applications). By late 1980 it was reported that several hundred such requests had been approved. Carter administration spokesmen defined the limits of such U.S. military cooperation with China in noting that "the United States and China seek neither a military alliance nor any joint defense planning" and that "the United States does not sell weapons to China." But they repeatedly implied that this policy could be subject to further change, especially if either or both countries faced "frontal assaults" on their common interests, presumably from the Soviet Union or its proxies.[7]

Although subject to differing interpretations, the Carter administration spokesmen's statements on the limits of the U.S.–China security relationship represented a departure from the past practice of the Administration and its two Republican predecessors: avoiding public explanations of the extent of Sino–U.S. security relations or their possible implications, and leaving the policy ambiguous. This approach was thought by some analysts to be a useful way to increase the impact of the developing Sino–U.S. ties on the Soviet Union; it presumably would prompt the Soviets to be more forthcoming and accommodating in their relations with the United States, in order to discourage Washington from developing even closer security ties with China.[8]

The ambiguity and uncertainty surrounding this aspect of American policy toward China continued following the inauguration of the Reagan Administration in January 1981, which received divided counsel on the sensitive question of military transfers to China. Accordingly, when Secretary Haig announced in Peking in June 1981 that the United States was now willing to consider the sale of weapons to China on a case-by-case basis, those Americans who favored sales portrayed the move as a welcome step forward, whereas those who opposed took some solace in the fact that the United States had still not agreed to sell any particular weapons to China.

POINTS OF AGREEMENT IN THE DEBATE

While arguments between proponents and opponents of increased military sales to China have ranged across a broad spectrum of issues, an analysis of the literature[9] and discussions with U.S. observers interested in U.S.–Chinese–Soviet relations[10] reveal a number of common themes and points of agreement. The consensus on these issues substantially narrows the scope of the debate, and in effect, provides a framework for it. Thus, for instance, the public warnings against U.S. military sales to China have sometimes been particularly shrill and unrealistic, with some commentaries warning of the dire consequences of a major American effort to "rearm" China. As seen in the discussion below, even advocates of closer U.S. military ties with China see insurmountable impediments to such efforts on both the U.S. and Chinese sides, thereby effectively eliminating them as realistic considerations in the current debate.

China

There is widespread agreement among U.S. specialists that China does not face an immediate crisis in its current military confrontation with the Soviet Union. The Chinese seem capable of resisting—albeit at great cost—a major Soviet conventional attack. And a Soviet nuclear attack against China is made less attractive because of the danger of a Chinese nuclear counterstrike against Soviet Asia. China is seen as clearly the weaker party militarily along the Sino–Soviet border, and its military capabilities are likely to decline relative to the Soviet Union if recent trends continue.[11]

While there is some disagreement on whether or not China has firmly decided on what kinds of weapons or other military equipment it would like to obtain from the United States or other foreign sources,[12] it seems clear that Chinese interest is quite selective. China has focused on obtaining the capability to produce such sophisticated equipment as fighter aircraft engines and radars, air-to-air missiles, and anti-tank missiles.[13] This accords with the views of many Western analysts who stress China's need for all-weather fighter aircraft, greater defensive capabilities against Soviet armor, and greater ability to provide air cover in nearby waters.[14]

There is general agreement among U.S. specialists that Chinese military forces are designed and deployed for the defense of China and are not well suited or positioned to project power far from China's borders. This does not mean that analysts agree that China would not use its existing forces to attack nearby areas. Indeed, some judge that prospects for a new Chinese attack against Vietnam remain high, and others remain quite skeptical of Chinese claims of "peaceful" intentions toward Taiwan.

There are also serious limits seen in Peking's ability to pay for and utilize large amounts of sophisticated weapons and weapons-related technology. China's recent economic programs actually have called for cuts in the defense budget, and they have given defense modernization the lowest priority among the four modernizations pursued by the present leadership.[15] China's difficulties in using the relatively sophisticated aircraft engine technology of the Soviet MiG-21 and the Rolls Royce "Spey" engine are seen as symptomatic of a broader Chinese technical weakness that was severely exacerbated by the disruption of Chinese higher education and research from 1966 to 1976.

Soviet Union

U.S. observers seem to agree generally that the Soviet Union will remain insecure about its position in the Far East for some time to come.[16] Even if China should alter its current foreign policy and move toward a more even-handed posture between the United States and the Soviet Union, Soviet planners are viewed as likely to remain distrustful of China and intent on maintaining Soviet military power in the region to protect against possible Chinese incursions. Military power is one of the few reliable sources of influence Moscow can bring to bear to protect its interests against a China that is growing in economic and military power and consolidating relations with the other two major powers in the region,

Japan and the United States. One of the major Soviet objectives in the region apparently is the protection and development of eastern Siberia, which Soviet planners reportedly view as an important element in their future economic development. A substantial Sino–Soviet reconciliation that would allow the Soviet Union to lower its guard against China is seen as only a very remote possibility for the foreseeable future.

Moscow's obvious desire to slow or halt the development of Sino–U.S. military ties is commonly perceived as motivated by concern over growing Chinese power and worry about a possibly emerging U.S.-backed global system, which would include China as well as the NATO countries and Japan, and which would be directed against the Soviet Union. It is thus repeatedly emphasized that Moscow will likely view U.S.–China military ties not solely in the context of their impact on Sino–Soviet military balance in Asia, but it will weigh their significance in the broader context of overall U.S. policy toward the Soviet Union.

United States

While the United States is seen as favoring a secure and independent China, a sharp change in China's recent defense policy toward a large-scale military buildup is widely viewed as contrary to U.S. interests. Such a move would upset the Chinese economic modernization program and perhaps would lead to political instability in China. It would upset the regional balance of power in East Asia in ways probably adverse to U.S. interests.

It is widely held that the United States would not likely offer free military assistance to China under current circumstances. Many observers accept the view that Chinese economic and technical weaknesses preclude large-scale U.S. military transfers that would fundamentally affect China's power projection capabilities over the near term. As a result, attention has focused on the political and symbolic repercussions of U.S.–China military ties. Ostentatious displays of U.S. military cooperation with China—if not backed by substance—are said to risk misinterpretation at home and abroad and should be avoided in favor of more quiet interchange. Several observers judge that under ideal conditions, it would be desirable to allow Sino–U.S. relations to mature for a few years before moving into the symbolically important area of military transfers.

A common recommendation in this regard is for more clear articulation and management of U.S. policy toward China. Several China specialists have noted, for instance, that Chinese leaders seemed to interpret past American behavior as indicating that the United States would soon be willing to transfer significant amounts of military equipment to China,[17] even though stated American policy at the time was to the contrary. These specialists added that the United States should avoid giving such impressions, unless it has the intention—and the political support at home—to follow through with the military sales. It is imperative, in their view, that the United States at some point "draw the line" with the Chinese, and thereby refuse Chinese pressure for military transfers without giving the Chinese the impression of a substantial decline in U.S. interest in close ties with China.

It is also widely held that increased U.S. military transfers to China would probably to some extent reduce U.S. influence and enhance Chinese influence elsewhere in Asia. They would prompt uncertainty among longstanding U.S. friends and allies in Asia, with the possible exceptions of Pakistan and Thailand, which might welcome such ties. They would deepen the suspicions of India, Taiwan and Vietnam toward the United States and would seriously reduce the prospect of a more cooperative U.S.–Soviet relationship.

Implications

These commonly held views suggest a narrower range of debate than appears at first glance and may make it easier for policy makers and other observers to assess realistically the pros and cons of U.S.–China military ties. For one thing, they indicate that the United States need not rush military transfers to China in order to help China against military threat from the Soviet Union. And, the United States need not worry excessively about a possible breakthrough in Sino–Soviet relations that would substantially reduce Soviet military preoccupations with China in Asia.

Also, China is likely to be quite selective in what it agrees to buy from the United States; and whatever the United States provides, it will probably take many years before it markedly increases China's power against its main adversary or increases China's ability to project power far beyond its borders.

Currently the main importance of U.S. transfers to China is a political or symbolic one. The transfers will clearly have an upsetting—though not necessarily adverse—effect on U.S. relations with other Asian states and the Soviet Union. Moscow will view such transfers not only in the context of its competition with China but also with an eye toward broad U.S. intentions toward the Soviet Union. Managing such a symbolically important and consequential relationship with China is often seen as requiring a more clearly defined perception of American interests in China that will avoid serious misinterpretation of American intentions at home and abroad.

STRATEGIC ASSUMPTIONS AND SCHOOLS OF THOUGHT

American observers have adopted different opinions and sided with different schools of thought on the question of U.S.–China military ties in large measure depending on their diverging assessments of the U.S.' foreign policy question—how far should the United States go in trying to improve relations with China in order to strengthen the U.S's international position against the Soviet Union? In particular, those who are deeply concerned with what they see as U.S. military "weakness" vis-a-vis the Soviet Union and who view China as strongly opposed to the Soviet Union and favorable to the United States tend to support increased U.S. military transfers to China. In contrast, those who are sanguine about U.S. power vis-a-vis the Soviet Union or who are skeptical of China's reliability or strength tend to oppose such sales.

DIFFERING STRATEGIC ASSUMPTIONS

Soviet Expansion and Its Challenge to the U.S.[18]

There is a large body of opinion in the United States that sees the Soviet Union as an expansionist power that seeks military, economic, and political preeminence in the world. (Some add that the Soviet Union is also anxious to project world revolution abroad in order to gain world ideological preeminence.) The Soviets are viewed as seeking to establish overall international superiority over the United States and are seen as on the ascendency in this effort. Current trends are viewed on balance as favoring Soviet ambitions, as present uncertainties and elements of instability in the world are believed to offer the Soviets fertile grounds for exploitation. Although Soviet interests have suffered certain setbacks, Soviet leaders have not been diverted from their ultimate objectives by momentary setbacks; they take a patient, long-term view of history, their confidence bolstered by a sense of destiny.

Another group emphasizes that even though Soviet leaders may strive for superiority and preeminence, circumstances unfavorable to the Soviet Union have profoundly affected their aspirations. Spokesmen of this group see the balance of world developments as frequently counter to Soviet designs. Domestic and international factors are perceived as limiting Soviet freedom to pursue their objectives. Soviet leaders are said to be aware of the fact that their ideological goals are presently unattainable and they are finding it necessary to adjust their policies to current realities. While the Soviet Union might strive to achieve superiority over the United

States, it is seen as having great difficulty in doing so. In short, the Soviet Union today qualifies only militarily as a superpower.

A third group believes that the Soviet Union is far less expansionistic than in the past and that Soviet policies are being shaped primarily by the country's requirements of internal social and economic development and a historic sense of insecurity. A more confident and secure Soviet leadership, according to this view, would be ready to act as a responsible participant in an increasingly interdependent world. Soviet military efforts are seen by this group as an unwanted drain on resources based on the Soviet view (perhaps misguided) of what is needed to maintain security. Since Soviet leaders understand that they are best served by international stability, their fundamental interests are reconcilable with those of the United States.

Observers who take this more benign view of Soviet policies and goals caution against looking at international developments in terms of superpower winners and losers. They judge that it is shortsighted automatically to equate Soviet setbacks with Western gains and vice versa. They stress that there are shared interests, dangers, and responsibilities between the superpowers in a number of areas. They advocate American policy aimed at increasing the Soviet stake in international stability and providing the Soviet Union with incentives for continued cooperation with the West.

U.S. Ability to Deal with Soviet Power

Closely related to the differing views of Soviet power and intentions are diverging assessments of U.S. ability to deal with that power. On one side are those who see American abilities as somewhat less than in the past, when the United States enjoyed dominant military, economic, and political influence in world affairs, but nonetheless still quite adequate to handle emerging Soviet power.

Others stress that U.S. power—especially military power—has declined steadily to a point where the Soviet Union now enjoys basic superiority over the United States in both conventional and strategic forces. They are generally optimistic, however, that the United States has begun to redress the military balance and will be able to catch up to the Soviet Union and close the so-called "window of vulnerability" by the end of the decade.[19]

Still others are more pessimistic about U.S. military capabilities, which they see as likely to decline further relative to the Soviet Union, short of a major shock in American foreign policy. It is sometimes noted that even though U.S. military capabilities may decline, the United States still has many more capable allies than does the Soviet Union; U.S. allies presumably can assist in checking Soviet expansion. However, several military planners and others have stressed that the allies have been slow and halting in their efforts to redress what is seen as an emerging East–West disequilibrium. They perceive, for instance, a weakened allied posture in Northeast Asia. They judge that the balance there has become more disadvantageous as U.S. forces have been drawn down as a result of the end of the Vietnam War and the diversion of forces to the Persian Gulf and Indian Ocean, and as the Soviet Union has continued to build steadily its air, naval, and ground forces in the area. They add that Japan—the largest and most capable U.S. ally in Asia—has helped inadequately to redress the balance over the past several years.[20]

China's Utility in U.S. Competition with the Soviet Union

China plays an important and helpful role in U.S. global and regional strategic planning, especially vis-a-vis the Soviet Union, according to many specialists.[21] They agree that China currently assists the United States by tying down Soviet troops and resources that otherwise might be focused against the West in Europe or the Middle East. This complicates Soviet defensive strategy, notably by raising Soviet uncertainty about the security of their Asian front should conflict break out farther west.[22]

But within this context, there are wide ranging views of China's utility in helping the United States in its competition with the Soviet Union. Some hold that China is a relatively weak source of leverage against the Soviet Union, so much so that it would be unlikely to cause serious concern for Moscow in the event of an East–West conflict elsewhere, or they judge that China's commitment to an anti-Soviet foreign policy is less strong than it might appear, emphasizing that China would be likely to arrange a *modus vivendi* with the Soviet Union rather than risk being drawn into a U.S.–Soviet confrontation.

Others see China as sufficiently strong militarily and reliable politically that it represents a useful partner for the United States in efforts to curb Soviet expansion. Still others stress that while China may be effective against the Soviets today, the growing disequilibrium between Chinese and Soviet forces along the border over the longer term could cause China to reconsider its anti-Soviet posture and reach an accommodation with Moscow contrary to U.S. interests.

CONTENDING SCHOOLS OF THOUGHT

Largely depending on how they assess the Soviet–U.S. rivalry and China's potential role in that rivalry, U.S. observers tend to identify with several discernible groups of opinion or schools of thought.[23] Individuals in these groups have naturally been inclined to give more stress to some issues, while soft-pedaling others. They have predictably done so with an eye toward safeguarding aspects of U.S. foreign policy of particular importance to them.

Thus, for example, many U.S. military and strategic planners both in and outside the Carter and Reagan Administrations have shown particular concern with what they have seen as the relative decline of U.S. military power vis-a-vis the Soviet Union in recent years. Dissatisfied with allied efforts to help redress the balance, they view U.S. military cooperation with China as a useful source of leverage that could help to remedy that decline.[24] In contrast, Americans interested in arms control with the Soviet Union frequently are concerned with restoring enough trust in U.S.–Soviet relations to facilitate conclusion of important agreements limiting strategic arms and theater nuclear forces. They see U.S. military moves toward China as contrary to this objective and as of marginal utility to the United States when compared to the importance of a major U.S.–Soviet arms accord. (It is worth noting that during the Carter Administration National Security Advisor Zbigniew Brzezinski seemed to favor the former view, while Secretary of State Cyrus Vance supported the latter.)

Soviet specialists are divided into two general groups on this issue. Some see Sino–U.S. military cooperation as contrary to what they judge should be the primary U.S. goal of establishing an international order based chiefly on a Soviet–American *modus vivendi*.[25] Many others, however, see the Soviet Union as a newly emerging great power and believe that the United States should work closely with other sources of world power—including China—in order to preclude more Soviet expansion and encourage the Soviets to adjust to and cooperate with the status quo. They see U.S. military cooperation with China as useful in this context.[26]

China specialists are also divided. Many are concerned with the negative impact a U.S. refusal to transfer military supplies would have on Sino–U.S. bilateral relations.[27] But many other Chinese specialists worry about potential negative consequences of closer military cooperation with China for future Sino–U.S. relations.[28]

Of course, not all views of U.S. arms transfers to China are governed by the Soviet–U.S.–Chinese triangular relationship. Thus, for example, many Asian specialists have reflected the uneasiness of the countries of the region over U.S.–China military cooperation.[29] Those with a particular interest in Taiwan have an obvious strong interest in blocking military ties with China. Meanwhile, Americans interested in increased trade with China have sometimes favored improved military ties as a means to show American good faith, to insure a fruitful economic relationship, and to build China's sense of security.[30]

PROS AND CONS OF U.S. MILITARY SALES TO CHINA

U.S. officials may well allow their particular views on the U.S.–Soviet–Chinese relationship or their special interests in U.S. foreign policy (e.g., Taiwan, trade with China, etc.) to influence their decisions with regard to U.S. military transfers to China. But they are unlikely to do so without careful review of the main advantages and disadvantages of such U.S. moves.

STRATEGIC CONSIDERATIONS

Some of the strongest arguments in favor of increased U.S. military transfers to China center on the effect they are said to have on Chinese and Soviet strategic planning. U.S. military transfers to China, it is asserted, will further insure that China would remain on the American side against the Soviet Union during the period of U.S. strategic vulnerability in the years ahead. The transfers would also serve to consolidate American ties with what is viewed as the emerging great power in Asia–China. A few planners point out that the U.S. military moves will help pave the way to what they see as a desirable and necessary Sino–U.S. mutual security arrangement in Northeast Asia – an alliance that they judge should also include Japan.[31]

U.S. military supplies would increase China's sense of security vis-à-vis the Soviet Union and reduce the chance that the latter would be able to intimidate or otherwise pressure China into a more pro-Soviet foreign policy stance. A greater Chinese sense of security is said to be necessary before the United States can expect China to join in serious discussions on limiting nuclear arms development.

While the U.S. transfers are not seen as substantially altering China's ability to project power against the Soviet Union, a number of analysts have said that such transfers could seriously complicate Soviet military plans in Asia. Moscow would not only have to devote more resources to countering whatever limited improvements are made in Chinese forces, but it would also have to worry more about conflict along its Asian front in the event of an East–West confrontation over Europe or the Middle East.

Some observers worry about the Soviet Union redeploying westward forces in Asia in the event of a crisis with the West in Europe. They add that such redeployments would be less likely under circumstances of closer Sino–American military cooperation. Closer military cooperation with China could also give the United States the option of using facilities in China (airfields, ports, etc.) in the event of a major confrontation with the

Soviets, thereby placing Soviet Central Asia and the Soviet Far East under greater pressure than at present.

Another strategic advantage seen by a few observers relates to Japan. Noting U.S. frustration with the slowness of the Japanese defense development and the seeming inability of the United States effectively to pressure Japan on this issue for fear of alienating our most important ally in Asia, these observers stress that closer military ties with China could help in these areas. They could allow the United States more latitude in pressuring the Japanese to do more in defense against the Soviet Union as well as in other areas, and could increase to some degree Japan's sense of vulnerability. This feeling of vulnerability is described as the most important element in influencing Japan to give defense a higher priority.

Opponents of U.S.–China military cooperation are not impressed by these supposed advantages and point to a variety of strategic disadvantages they see associated with U.S.–China military ties. For one thing, China's current nuclear strategic capability is potentially threatening to the United States, and U.S. aid to China's conventional forces could presumably allow China to devote more attention to developing strategic weapons. China is said to be likely over the longer term to pose a threat to U.S. interests in Asia. As it gets stronger with U.S. support, China may well act more independently and assertively and come into conflict with some of its neighbors whose interests are close to those of the United States—Taiwan is the most obvious area for such conflict.[32]

Closer military cooperation with China could lead to such negative consequences as even more Soviet military pressure on China[33] or perhaps a punitive Soviet strike along the Chinese border.[34] The Soviet countermoves, if successful, could undercut the relatively pro-Western leadership of China and discredit those Chinese officials who have linked China's defensive strategy to close association with the United States.[35]

The Soviet Union could react to U.S. transfers of military supplies to China with countermeasures involving Vietnam or possibly India, perhaps including stepped-up efforts to establish Soviet military installations in Southeast and South Asia. These moves not only would help the Soviet Union to encircle China and curb Chinese influence in Asia, but they would also seriously challenge the ability of the United States to defend sea lines of communication in these important areas. U.S. moves toward military ties with China could seriously dampen Soviet interest in arms control with the United States and upset existing East–West understandings for these negotiations. In particular, Moscow would be more likely to demand compensation in any disarmament proposal due to Chinese nuclear forces—something the United States has rejected in the past.

POLITICAL–ECONOMIC FACTORS

Perhaps the strongest political argument for going ahead with arms sales to China is that U.S. leaders have already given Chinese officials the impression that significant sales would be allowed. To reverse course at this stage could lead to serious complications in Sino–U.S. relations. The transfers would show American "good faith," build support for and establish U.S. influence with the relatively pragmatic leaders currently governing China, and promote important channels of communications with segments of the Chinese military leadership who might otherwise remain skeptical of China's recent tilt toward the United States.[36] It is asserted that preparing such a solid foundation for Sino–U.S. ties is essential in order to permit the relationship to withstand future difficulties over such issues as Taiwan,[37] U.S.–Soviet arms control, and human rights.

Military transfers to China are also said to provide the United States with a "China card" useful in promoting more positive Soviet behavior toward the United States or in compensating the United States for Soviet gains made elsewhere in the Third World. They reportedly have indirect advantages for U.S. trade with China, as China is said to be likely to give business people associated with its major military backer more advantageous treatment than their competitors from other countries. Closer military ties with China could reduce U.S. dependency on Japan and increase U.S. leverage over the Japanese on a variety of issues, including U.S.–Japanese trade and defense disagreements.[38]

Opponents of military transfers raise a host of possible political and economic disadvantages to such ties. They could lock the United States into an anti-Soviet posture in international affairs at a time when the United States may have more to gain from cooperation than confrontation with the Soviets. This could happen even if the United States wanted only a limited military relationship with China, because such relationships, once started, develop rapidly and prove difficult to stop.

The transfers could link the United States closely with the Chinese side of the Sino–Soviet dispute in Asia, notably reducing prospects for more independent U.S. policies vis-a-vis Vietnam or India. They could promote a view that the United States perceives China, rather than Japan, as its main ally in Asia, thereby leading to an erosion of the U.S.–Japan Alliance. They could signal a loss of U.S. influence in Asia, as well as a loss in influence over China's future behavior. The latter whould be even more likely if the United States agreed to coproduction-type arrangements with China that would allow the Chinese to have full control over the use of weapons produced with technology supplied by the United States.[39]

In view of China's history of political instability, it is quite possible that a new leadership less favorable to the United States could emerge in China over the next few years. The example of Soviet military cooperation with China in the 1950s is also not reassuring, as it seemed to prompt unrealistic Chinese expectations that the Soviet Union was unwilling to fulfill. This led to a serious downturn in relations—a pattern which could be followed in Sino–American relations during the 1980s if the Chinese military leaders and other officials came to rely too heavily on American supplies and support.

China is also seen as gaining much more than it gives in its new relationship with the United States. Some stress the contrast between the risks the United States would take in increasing military ties with China and the fact that the Americans would still have no guarantee that China would be any more likely to side with the United States on issues important to U.S. interests vis-a-vis the Soviet Union or elsewhere.[40]

PROSPECTS AND OPTIONS

If and when China and the United States can settle amicably their differences over U.S. arms sales and other relations with Taiwan, it appears likely that U.S. officials in the Reagan Administration, like their counterparts in the Carter government, will be buffeted by cross-currents of political opinion as they attempt to deal with the issue of increased U.S. military supplies to China. Trends in U.S.–Soviet–Chinese relations still seem to favor those Americans who wish to transfer more military equipment and arms to China. Support for such sales is likely to build so long as U.S. relations with the Soviet Union remain characterized more by hostility than by cooperation, U.S. relations with China evolve toward closer friendship, and Sino–Soviet relations remain stalemated.

The Chinese appear likely to be receptive to U.S. offers of more advanced military equipment and technology. A flat U.S. refusal to sell such material to China would almost certainly prompt expressions of strong disappointment by Chinese leaders, who would see it as a clear sign of America's lack of trust of China and U.S. determination to maintain an "arms embargo" against them. When taken together with the Reagan Administration's well-known differences with Peking over U.S.–Taiwan relations, the refusal could lead to a serious downturn in Sino–American relations.

The main worry about the military transfers concerns their symbolic importance, especially their implications for future U.S. policy toward China and their meaning for American relations with the Soviet Union,

Japan, and other Asian states. Here lies what many see as a major challenge to American policy makers in dealing with China.

The incrementally developing American security relationship with China over the past decade has from time to time led to misconceptions of U.S. intentions on the part of domestic interest groups, the Soviet Union, Asian countries, and even the Chinese leaders. If U.S. leaders maintain a vagueness about China's place in Washington's security policy while moving ahead with more military transfers to China, various interested parties could make too extreme an interpretation of U.S. objectives; their reactions could well be contrary to U.S. interests. Thus, Chinese leaders might incorrectly see such transfers as signaling a major strengthening of what they may view as Washington's commitment to protect China's security against the Soviet Union. Soviet leaders might perceive them as the consumation of a de facto U.S.–China military alliance that must be actively resisted by the Soviet Union. Japan and other U.S. allies and friends in Asia could see them as signaling a fundamental shift in U.S. interests in Asia, away from them but toward China as the main backer of U.S. security interests in the region.

By contrast, a different approach, which would allow the United States to make clearer to the various interested parties the limited objectives of its China policy, could reduce possible adverse consequences for U.S. interests. Under these circumstances, the United States would agree to increase to a carefully limited degree U.S. military transfers to China, including some defensive weapons or related technology. It would use the opportunity to serve notice to Peking that the Chinese leaders should expect no more such military help until Sino–American relations have matured over several more years. This policy would remain in effect barring a gross change in the international balance of power or a substantial increase in Soviet military power in Asia designed to pressure China into a neutral or anti-American posture. The United States could continue to solidify its ties with China in economic and political areas that are seen as having little negative consequence for U.S. interests.

It would be particularly important to clarify the limits of Sino–American military cooperation to the leaders of the Soviet Union. It could be noted in conversations with the Soviets that U.S. supplies to China are in part governed by a desire to maintain, but not to narrow substantially, the current gap in the military capabilities of Sino–Soviet forces along China's northern border. Thus, Moscow would know that any major Soviet effort to expand its military power in Asia in order to intimidate China would likely prompt increased U.S. support for China's military modernization.

Japan and other Asian states would also have to be reassured. A solid consensus in the United States in support of the limited military relationship with China would appear to be required. Such a consensus might prove difficult to build, and yet remain within the confines of the limited military relationship noted above. In particular, some in Congress may be inclined to ask for a "quid pro quo" for U.S. military help to China. While this could involve preferential treatment of U.S. business representatives in trade with China, some U.S. representatives might demand that the United States be compensated with increased military access to China, such as with basing rights. Of course, such a move would be more likely to increase the risks of Soviet and other foreign reactions contrary to U.S. interests.

Another option is to use limited military transfers as the first in a series of steps leading to some sort of mutual security arrangement with China against the Soviet Union. These steps could include U.S. training of Chinese military personnel, the provision of large amounts of U.S. military aid, the stationing of American military experts in China, the widespread presence of U.S. intelligence facilities in China, Sino–American maneuvers and exercises, and U.S. planning with Chinese forces. While such an approach would clearly have at least a short-term positive impact on U.S.–China relations, it would make almost impossible any lasting U.S. reassurance of the Soviet Union and other interested Asian nations over American intentions toward China. Should U.S. planners pursue this option, they probably should continue an ambiguous and secretive policy with regard to U.S.–China security ties. As noted above, such a policy carries the risk of prompting extreme international reactions—reactions that could seriously complicate American interests both at home and abroad.

In short, U.S. refusal to sell any weapons or related technology to China could seriously affect Sino–American relations and, by extension, a variety of important U.S. interests in world affairs. But the risks of moving ahead with such sales—whether or not they are designed to foster a close Sino–American mutual security arrangement against the Soviet Union—will remain great so long as U.S. leaders remain vague about the objectives of their strategic relationship with China. The risks of a carefully limited U.S. military transfer to China could be reduced substantially if U.S. leaders shifted to a more clearly defined policy of moderate U.S. objectives that could assure friends and foes alike of American intentions toward China.

ASSESSMENT

The trend in the U.S.–Soviet–Chinese triangular relationship after 3 years of the Reagan Administration still appears to argue for the United

States to seek closer strategic cooperation with China against the Soviet Union, including eventually the sale of U.S. military equipment to China. However, developments in the relationship have made it even more likely than before that the United States will adopt a cautious approach to military cooperation—a stance likely to be reciprocated by Peking. Thus, for example, the strong Sino–U.S. differences over U.S. arms sales to Taiwan have had some important effects on U.S. assessments of China. For one thing, Peking's tough stance brought to a halt the incremental forward movement in U.S. military cooperation with China, giving time for U.S. specialists to examine more carefully the pros and cons of the issue, and for opponents to marshall arguments against such interchanges with China.

It also upset the calculations of some Americans that China was so anxious for military cooperation with the United States against the Soviet Union, including U.S. arms sales to China, that it would be willing to soft-pedal sensitive military ties. As a result, these Americans were surprised by Peking's strong stance on Taiwan and freeze on discussion of military sale with the United States; they were forced to reassess their views on China's potential utility in cooperating with the United States against the Soviet Union.

Such reassessment has become more urgent as it has become clearer during the past 2 years that Peking is shifting toward a more independent posture in foreign affairs that has involved closer alignment with third world countries against the superpowers—an allusion meant to include both the Soviet Union and the United States. Chinese rhetoric softpedaled past invective against the international "menace" posed by Soviet expansionism; sometimes saw the Soviet Union as at least temporarily bogged down by difficulties in the domestic economy, leadership succession, and international involvement in such trouble spots as Poland, Afghanistan, and Indochina; and showed new interest in at least some improvement in bilateral relations. Meanwhile, Chinese media were more critical of U.S. policies, not only concerning Taiwan but also Korea, Latin America, the Middle East, and elsewhere. Chinese leaders dropped their past calls for the formation of an international united front against "hegemony," that in the past had been directed at marshalling international forces, including the United States, against Soviet expansion; and they no longer called for the development of a long-term strategic relationship with the United States, as they had in the past.

Nevertheless, there still appeared to be strict limits as to how far China is prepared to go in easing tensions with the Soviet Union or in distancing itself from the United States. For one thing, China remains preoccupied with the strategic problem of dealing with continued Soviet efforts to use military power and influence in Asia to curb and contain Chinese interests and influence. This policy has been fundamental to Chinese foreign policy ever since the Sino–Soviet border clashes of 1969, when the Soviet Union showed their "true colors" to China by threatening to attack Chinese nuclear installations and by launching the Asian Collective Security System—a thinly veiled Soviet effort to foster Soviet-led Asian containment of China.

At present, the manifestations of this Soviet approach are seen in the steady improvement of Soviet forces along the Sino–Soviet and Sino–Mongolian frontiers, the gradual buildup of Soviet naval and air power in the Western Pacific, and the growth and consolidation of Soviet influence in Southwest and Southeast Asia, via Afghanistan and Vietnam. Although China recently may view the Soviet Union as bogged down in a variety of unproductive commitments in these and other areas, it has seen no sign of any fundamental shift in Moscow's perceived desire to practice hegemony over China. Peking has also taken due note of the potential for internal difficulties in the Soviet Union, caused by political succession and economic setbacks, but it has perceived little prospect that these events will cause the basic reorientation in Soviet foreign policy that would be needed before a major Sino–Soviet reconciliation would appear possible.

As a result, ties with the United States and the West will remain an important element in Chinese foreign policy. At the same time, while China might expect to derive some economic benefit from improved trade and exchanges with the Soviet Union, the United States, Japan and West Europe have the technology, equipment, and know-how that China will need to make progress toward economic modernization by the end of the century.

Meanwhile, Americans remain willing to accommodate interests elsewhere for the sake of maintaining and developing good strategic relations with China. This has been seen most recently in the Sino–U.S. joint communiqué on August 17, 1982, whereby the United States agreed to qualitative and quantitative limits to its arms sales to Taiwan and implied a willingness to curb gradually such sales over the next few years, rather than risk a downgrading in U.S. relations with the PRC. As in the past, the motivating force behind American policy of reconciliation with China has been the search for strategic advantage.

The attempt to deal with Soviet power is likely to preoccupy American leaders for some years to come. Because of the perceived inability of the United States to tackle the Soviet Union on its own, Americans will likely try to elicit increased support from U.S. allies, which, like the United States, remain influenced by ideological, economic, and political trends that undercut a firm defense posture against the Soviet Union. Thus, Washington will have to cast its net wider in the search for anti-Soviet leverage, assuring China a continuing important role in future U.S. strategic calculations, provided she remains cool to the Soviet Union. In this context, the United States can be expected to accommodate its interests on secondary considerations in order to maintain and develop improved relations with China in the interests of strengthening its international effort to deal with the problem of Soviet expansionism.

NOTES

1. For background, see Richard Solomon (ed.) *The China Factor* (Englewood Cliffs, N.J.: Prentice-Hall, 1981).

2. For more information see:

A. Doak Barnett, *U.S. Arms Sales: The China–Taiwan Tangle*, Washington, D.C. The Brookings Institution. 1982.

U.S. Congress, Committee on Foreign Affairs, Subcommittee on Asian and Pacific Affairs, *The New Era in East Asia*, Washington, D.C.: U.S. Govt. Printing Office, 1981.

– – –. *The United States and the People's Republic of China: Issues for the 1980s*, Washington, D.C. Govt. Printing Office, 1980.

U.S. Congress. Senate. Committee on Foreign Relations. *The Implications of U.S.–China Military Cooperation.* Washington, D.C. Govt. Printing Office, 1982

U.S. Library of Congress. Congressional Research Service. *Increased U.S. Military Sales to China: Arguments and Alternatives.* Washington, D.C. May 20, 1981.

3. See Barnett, *U.S. Arms Sales.* See also reports in the *Far Eastern Economic Review*, especially those by Richard Nations and Nayan Chanda, during 1981–1982.

4. For further information and background, see U.S. Library of Congress. Congressional Research Service. *China–U.S. Relations.* Issue Brief IB 76053 (periodically updated).

5. *Ibid.*

6. See notes 1 and 2 as well as the following selected readings on this issue (see full titles in References at the end of the chapter): A. Doak Barnett (1977); B.D.M. Corporation (1976); Roger Brown (1976); John M. Collins (1980); Defense Intelligence Agency (1976); Angus M. Fraser (1979); Banning Garrett (1979a, 1979b, 1979c); Thomas Gottlieb (1977); International Institute for Strategic Studies (1981, 1980); Joseph Kraft (1978); Leo Yueh-yun Liu (1980); Edward Luttwak (1978); Thomas A. Marks (1980); Drew Middleton (1978); Lelands Ness (1978); Robert L. Pfaltzgraff, Jr. (1980); Michael Pillsbury (1977, 1975); Clarence A. Robinson (1980); Francis J. Romance (1980); William Schneider (1979); Richard H. Solomon (1979); Ross Terrill (1977, 1978); Allen S. Whiting (1980); Robert L. Worden (1980); Peter L. Young (1979); and the various relevant reports by the U.S. Congess listed in the bibliography at the end of the present book.

7. For additional information, see U.S. Department of State. GIST. *U.S.–China Security Relationship.* Washington, July 1980.

8. Others, however, saw the ambiguity surrounding U.S.-China relations as designed to hide the absence of a well coordinated U.S. policy toward China that could be explained clearly and could receive the full support of the American people and their representatives in Congress.

9. See sources in note 6.

10. Over the past 2 years, the author has held more than 50 interviews with U.S. Government officials, academic experts, journalists, business representatives, and others knowledgeable about the issue of U.S. military relations with China. Given the institutional sensitivities that surround this subject, individuals with whom the author spoke were assured that their comments would not be for attribution.

11. For a good summary of recent opinion on Chinese military capabilities, see U.S. Congress, Senate, Committee on Foreign Affairs. *The Implications of U.S.-China Military Cooperation.*

12. In this regard, several military specialists have pointed to the seeming inconsistency in Peking's defense procurement process, noting in particular the on-again, off-again Chinese interest in the British "Harrier" aircraft. China also was reported very interested in French antitank missiles, but agreements on this item have not been reached. The reasons given for this seeming Chinese inconsistency vary: Some point to China's recently heightened awareness of its inability to pay for foreign equipment. Others note the reluctance of some western countries to offend the Soviet Union by being one of the first Western powers to sell weapons to China. Still others emphasize that Western countries are unwilling to supply China with sophisticated military equipment that they judge would be of little practical use for China given the current, relatively low technical competence in that country.

13. Chinese leaders were seen as pressing the United States to permit such military transfers, both to enhance China's military power against the Soviet Union and to demonstrate America's growing commitment to China's defense. Thus, for example, Su Yu, a senior Chinese military leader, told the Japanese press in March 1980 that America's willingness to move ahead with sales of military equipment to China was seen by the Chinese as a key indication of the U.S. commitment to work with China in a common front against the Soviet Union. Vice Premier Geng Biao and Vice Foreign Minister Zhang Wenjin both subsequently told the Western news media that they anticipate that the United States will sell arms to China in the future. Evidence of this view was seen in the *Washington Post*, November 23, 1980.

14. Some of the best work in this area is seen in the recent writings of Douglas T. Stuart and William T. Tow (eds.), *China, the Soviet Union, and the West* (Boulder, Colo.: Westview, 1982).

15. While there was some public debate in China during 1977 and 1978 over how much of China's resources would be devoted to modernizing the defense establishment, the trend since then has been to give defense modernization a low priority.

16. See reaction to remarks by William Hyland in U.S. Congress. Senate. Committee on Foreign Relations. *The Implications of U.S.-China Military Cooperation. op. cit.*

17. See discussion of this subject, *ibid.*

18. For a useful wrap-up on U.S. views of Soviet power and U.S. ability to deal with Soviet power see U.S. Library of Congress. Congressional Research Service. *Soviet Strategic Objectives and SALT II: American Perceptions.* Report No. 78-119F.

19. They sometimes add that projected Soviet economic and other internal difficulties by the end of the decade will restrict Moscow's ability to expand its military power as in the past and will thereby enhance America's ability to close the power gap.

Technically speaking, the term *window of vulnerability* refers to the period when U.S. land-based ICBMs will be vulnerable to Soviet attack—a period that is supposed to end with the deployment of the new MX missile. However, the term is sometimes used more broadly to refer to what is seen as a general superpower imbalance, against the United States.

20. Some analysts offer similarly alarming views of the decline of the allied position in the face of Soviet or Soviet-backed expansionism in Southeast Asia, and they argue strongly for closer relations with China as a means to sustain a favorable equilibrium. As in the case of Northeast Asia, these analysts see it as fundamentally important that the U.S. increase military transfers to China, not so much because they will increase Chinese power against

the Soviet Union, but because they will consolidate U.S. relations with China and thereby increase prospects that China will side with the United States in the event of East–West confrontation in Northeast or Southeast Asia.

In contrast, some other military planners are less concerned by the changing military balance in East Asia. They tend to emphasize Soviet logistical problems and other sources of military vulnerability as well as Moscow's political isolation. In particular, Moscow has made few gains in expanding its influence as the United States has pulled back its forces, and the two major regional powers–China and Japan–continue to work against Soviet interests in the region.

21. See the review in U.S. Library of Congress. Congressional Research Service. *Increased U.S. Military Transfers to China: Arguments and Alternatives.* op. cit.

22. Soviet military planners are thought to assess China's limited military capabilities much more realistically than the shrill Soviet media commentaries that warn of the China "threat." Nevertheless, Soviet logistical problems and other weakness in Asia as well as their assessment of the probable cost of any protracted conflict with China are thought to promote general Soviet concern about the Asian front.

23. These schools of thought are by no means clearly defined groups, with uniform points of view. Rather, they represent only the beginnings of the development of focal points in the debate on U.S.–China military ties.

24. This point of view was voiced by a number of U.S. military analysts who were interviewed, although they acknowledged that public attention to their approach has been sparse. For a classic example of this point of view, see the articles by Michael Pillsbury cited in note 6.

It should be added that these military planners tend to stress that even though Chinese military capabilities are not expected to increase rapidly as a result of increased U.S. military transfers to China, closer U.S.–China security ties are beneficial to the United States. They would help ensure that China would remain on the U.S. side in the event of a major East–West confrontation and would increase Soviet worry about China taking action against the Soviet Union in conjunction with U.S. actions against the Soviet Union.

25. See in particular, the testimony of Raymond Garthoff of the Brookings Institution before the House Foreign Affairs Committee on August 26, 1980. Marshall Shulman, a senior State Department adviser during the Carter Administration, has also been a strong proponent of this point of view.

26. Harvard University Professor and Reagan Administration adviser Richard Pipes has been outspoken in this regard.

27. Several judge that the United States might seriously disappoint the Chinese leaders by not following through with military supplies, after having given the Chinese the impression during visits and other interchanges that such equipment would be forthcoming. Some of them add that the supply of limited amounts of weapons and weapons-related technology represents an effective way to consolidate relations with the Chinese leadership. Amicable Sino–American relations are seen as a useful means to stabilize the situation in Asia in the face of possible internal and external challenges, including possible Soviet expansion.

While this view has been held by several important China analysts in the U.S. Government in recent years, it has not been subject to much media attention. For a variation on this view, see Roger Brown's article, cited in note 6.

28. Some worry about leadership instability in China or voice concern over Chinese intentions toward their neighbors. U.S. military ties might identify the United States too closely with only one group in the Chinese leadership–a group whose tenure may be limited and whose successors may not be favorably disposed to the United States.

Some analysts, who have been critical of recent developments in China, have been associated with this view. See in particular Ray Cline's analysis of the "China Card" in the *Washington Star*, October 14, 1980.

Other China specialists stress that the Chinese may come to rely too heavily on the United States, or may find U.S. military equipment inappropriate for China's military modernization—developments possibly leading to a severe downturn in U.S.–Chinese relations in the future. A. Doak Barnett and Allen Whiting voiced these kinds of reservations in testimony before the House Foreign Affairs Committee on July 22, 1980.

29. Their concerns focus on China's irredentist claims and its potential role as a destabilizing force in the region—factors that are seen as possibly more difficult to deal with if the United States seems to defer more to China's interests in Asian affairs.

Stanford University Professor Harry Harding has urged that there be closer understanding between the United States and its allies in Asia on China policy before the United States begins to sell weapons or weapons-related technology to China. See his testimony before the House Foreign Affairs Committee, September 25, 1980.

30. These developments reportedly will increase the likelihood that China will play a stabilizing rather than disruptive role in the economically important East Asian region. This view was voiced by some U.S. business persons with a special interest in East Asian affairs.

31. These analysts are predictably vague in defining the outlines of this proposed mutual security arrangement, although they repeatedly compare it directly with the North Atlantic alliance.

32. For instance, it is claimed that U.S. military transfers to China might so increase China's perceived leverage over the United States in Asia that Peking might take more forceful actions—such as a naval blockade—against the island. U.S. arms are not thought to increase substantially the Chinese ability to conduct such an operation, at least for several years. But the transfers could signal Peking, Taiwan, and Taiwan's leading trading partners of a shift in U.S. priorities in the region, suggesting that the United States would not react strongly to such a Chinese resort to force against the island. In the event of a Chinese blockade—which is thought by many to be within the range of China's current military capabilities—the United States would have to decide whether to confront its strategically important friend in Asia or allow Taiwan to be pressured into an accomodation with the mainland.

33. Some analysts would welcome an increase in Soviet deployments against China because they would reduce Soviet ability to confront the West elsewhere.

34. While a few analysts see Moscow as possibly reacting immediately with force to U.S. sales to China, others stress that Soviet military planners have not been rash in their military actions against China. Several have added that the United States should be wary that even though Soviet planners may appear cautious, U.S. military cooperation could quickly build to a point where it would cross an ill-defined "threshold" of Soviet tolerance, leading to a harsh military response against China, and possibly, the United States.

35. Some China specialists argue that China is unlikely to change its anti-Soviet posture and will continue to tie down Soviet forces, whether or not the United States increases military transfers to China. They therefore see little need for the United States to risk the possible disadvantages of closer U.S.–China military ties.

36. It has been noted by some that Chinese military leaders have received less benefit than other Chinese leaders from China's modernization program or its opening to the West; they therefore would have less to lose, and possibly more to gain, if that program and China's pragmatic, relatively pro-Western leadership under Deng Xiaoping were changed.

37. It was said by some at the time of the end of the Carter Administration and the start of the Reagan Administration, that the United States cannot expect relations with China to remain cordial if it continued—as expected—to sell military equipment to Taiwan while maintaining a de facto arms embargo against the People's Republic. In particular, some said that the United States could avoid a downturn in U.S.–China relations and still go ahead with the proposed sale of the FX fighter aircraft to Taiwan by simultaneously allowing the

transfer to China of sophisticated U.S. aircraft engine technology needed for the Chinese fighter aircraft program.

38. A few observers have claimed that Japan's leaders have been privately arguing against U.S. arms sales to China in part for selfish reasons. The Japanese alledgedly judge that if the United States moves ahead with such sales, China will favor U.S. businessmen over Japanese businessmen on potentially lucrative trade deals.

39. Even some analysts who favor arms transfers to China argue that such sales should be seen only as a supplement to, not a substitute for, U.S. power in the region. Otherwise, they warn, the United States would become too dependent on China to protect U.S. interests in the region. However, others favoring Sino–American security ties judge that it is unrealistic to expect the United States not to use China as a substitute for U.S. power in East Asia to some degree. They emphasize that the United States needs to consolidate its forces in order to deal with the Soviet Union in other important areas, notably the Persian Gulf, and that it should use China as a strategic bulwark in East Asia. Some note that the United States in effect has already started this kind of approach in Southeast Asia, where China–and not the United States–is seen as the main strategic guarantor of American interests in Thailand against military pressure from Soviet-backed Vietnam.

Meanwhile, several observers have pointed out that many of the disadvantages of U.S. arms sales to China could be overcome if West Europeans–not Americans–sold arms to China. But others have added that these countries frequently have shown themselves to be very sensitive to Soviet pressure not to sell arms to China–pressure that would presumably have less effect on the United States.

40. Some analysts claim that Chinese leaders lack a fixed and viable defense strategy and that U.S. military supplies could prove to be less than fully useful if Chinese defense plans changed in the future. Chinese leaders might then be inclined to blame the United States for transferring "inappropriate" equipment at great cost to China's limited economic resources.

REFERENCES

Barnett, A. D. "Military Security Relations between China and the United States." *Foreign Affairs*, April 1977.

——. *U.S. Arms Sales: The China–Taiwan Tangle.* Washington, D.C.: The Brookings Institution, 1982.

BDM Corporation. *U.S.–PRC–USSR Triangle: An Analysis of Options for Post-Mao China.* Vienna, Va.: BDM Corp., 1976.

Brown, R. G. "Chinese Politics and American Policy: A New Look at the Triangle." *Foreign Policy*, Summer 1976.

Cline, R. "China Card" *The Washington Star*, October 14, 1980.

Collins, J. M. *U.S.–Soviet Military Balance, 1960–1980.* New York: McGraw-Hill, 1980.

Defense Intelligence Agency. *Handbook of the Chinese Armed Forces*. July 1976.

Fraser, A. M. "Military Modernization in China." *Problems of Communism*, September/December 1979.

Garret, B., "The China Card: To Play or not to Play." *Contemporary China*, Spring 1979. (a)

———. "China Policy and the Strategic Triangle." In *Eagle Entangled American Foreign Policy in a Complex World*. New York: Longman, 1979. (b)

———. "A Wild Card in the Global Balance." *Internews*, April 9, 1979. (c)

Gottlieb, T., *Chinese Foreign Policy Factionalism and the Origins of the Strategic Triangle*. Santa Monica, Calif.: Rand R-1902-NA, November 1977.

International Institute for Strategic Studies. *Military Balance, 1981–1982*. London, 1981.

———. *Strategic Survey, 1979*. London, 1980.

Kraft, J. "Triangle Diplomacy Without Magic." *Washington Post*, May 25, 1978.

Liu, L. Y. Y. "The Modernization of the Chinese Military." *Current History*, September 1980.

Luttwak, E. "Against the China Card." *Commentary*, October 1978.

Marks, T. A. "The Chinese Road to Military Modernization—And a U.S. Dilemma." *Strategic Review*, Summer 1980.

Middleton, D. *The Duel of the Giants: China and Russia in Asia*. New York: Charles Scribner's Sons, 1978.

Ness, L. "Chinese Army." *Armies and Weapons*, June 1978.

Pfaltzgraff, R. L. "China, Soviet Strategy, and American Policy." *International Security*, Spring 1980.

Pillsbury, M. "Future Sino–American Security Ties: The View From Tokyo, Moscow, and Peking." *International Security*, Spring 1977.

———. "U.S. Chinese Military Ties." *Foreign Policy*, Fall 1975.

Robinson, C. "China's Technology Impresses Visitors." *Aviation Week and Space Technology*, October 6, 1980.

Romance, F. J. "Modernization of China's Armed Forces." *Asian Survey*, March 1980.

Schneider, William "China's Military Power." In *About Face: The China Decision and its Consequences*. New Rochelle, N.Y.: Arlington House.

Solomon, R. H. (ed.) *The China Factor: Sino–American Relations and the Global Scene*. Englewood Cliffs, N.J.: Prentice-Hall, 1981.

———. (ed.) *Asian Security in the 1980's: Problems and Policies for a Time of Transition*. Santa Monica, Calif.: Rand Corporation, R-2492-ISA, November 1979.

Stuart, D. T. and W. T. Tow (eds.) *China, The Soviet Union and the West*. Boulder, Colo.: Westview, 1982.

Terrill, R. "China and the World: Self-Reliance or Interdependence." *Foreign Affairs*, January 1977.

———. "The Strategy of the Chinese Card." *Asian Wall Street Journal*, July 20, 1978.

U.S. Congress. Committee on Foreign Affairs. Subcommittee on Asian and Pacific Affairs. *The New Era in East Asia*. Washington, D.C.: U.S. Government Printing Office,

———. U.S. Congress. *The United States and the People's Republic of China: Issues for the 1980s*. Washington, D.C.: U.S. Government Printing Office, 1980.

———. House. Committee on Foreign Affairs. Subcommittee on Asian and Pacific Affairs. *Playing the China Card: Implications for United States–Chinese–Soviet Relations*. Washington, D.C.: U.S. Government Printing Office, 1979.

———. *Normalization of Relations with the People's Republic of China: Practical Implications*. Hearings, 95th Congress, 1st Session. Washington, D.C.: U.S. Government Printing Office, 1977.

———. House. Special Subcommittee on Investigation. *United States–China Relations: The Process of Normalization of Relations*. Hearings, 94th Congress, 1st and 2nd Sessions. Washington, D.C.: U.S. Government Printing Office, 1976.

———. House. Special Subcommittee on Investigations. *United States–China Relations: The Process of Normalization of Relations*. Hearings, 94th Congress, 1st and 2nd Sessions. Washington, D.C.: U.S. Government Printing Office, 1976.

———. Subcommittee on Future Foreign Policy Research and Development. *United States–Soviet–China: The Great Power Triangle*. Washington, D.C.: U.S. Government Printing Office, 1976.

———. House. Committee on International Relations. *United States–Soviet Union–China: The Great Power Triangle, Summary of Hearings conducted by the Subcommittee on Future Foreign Policy Research and Development of the Committee on International Relations*. Washington, D.C.: U.S. Government Printing Office, 1977.

———. Senate. Committee on Foreign Relations. *Sino–American Relations: A New Turn*. Washington, D.C.: U.S. Government Printing Office, 1979.

———. *The Implications of U.S.–China Military Cooperation*. Washington, D.C.: U.S. Government Printing Office, 1982.

U.S. Department of State. *U.S.–China Security Relationship*. Gist. Washington, D.C.: U.S. Government Printing Office, 1980.

U.S. Library of Congress. Congressional Research Service. *Soviet Strategic Objectives and SALT II: American Perceptions*. Report No. 78-119F, 1981.

———. *Increased U.S. Military Sales to China: Arguments and Alternatives*. Washington, D.C., May 20, 1981.

———. *China–U.S. Relations*. Issue Brief IB 76053, periodically updated.

Whiting, A. S. "China and the Superpowers: Toward the Year 2000." *Daedalus*, Fall 1980.

Worden, R. L. *Chinese Militia in Evolution*. Defense Intelligence Agency DDB-1100-285-80, May 1980.

Young, P. L. "China's Military Capabilities." *Asian Defense Journal*, February 1979.

4

Emergence of an "Independent" Chinese Foreign Policy and Shifts in Sino–U.S. Relations

Carol Lee Hamrin

The most noticeable variable in U.S.–China relations since the start of 1981 is the distinct emergence of an "independent" Chinese foreign policy posture, placing the United States and the Soviet Union on more or less an equal footing very similar to the Chinese position before 1978. In mid-April 1982, Premier Zhao Ziyang modified China's formal line on foreign affairs to stress that "strengthening Third World unity" was as important as "opposing hegemonism and safeguarding world peace." This elevation of Peking's third world concerns had been foreshadowed by the Premier at the North–South Conference in Cancun, Mexico, during the fall of 1981, when he offered a new level of Chinese support for moderate third world initiatives to build a new international economic order.[1] He also had stressed the inseparability of the maintenance of peace and the promotion of development. In 1982, Chinese leaders also began attributing a new importance to South–South consultation and assistance as a necessary means of promoting effective North–South dialogue.

A surprisingly even-handed critique of both superpowers for their contribution to international instability emerged in tandem with this third world emphasis. In mid-April, Zhao told a visiting Somali official:

> Facts have shown time and again that the superpowers are bent on controlling, subverting, exploiting and invading other countries, Third World countries in particular, and thus have posed a grave threat to peace and tranquility in the world. As an African saying goes, "when elephants fight each other, grass suffers."[2]

This chapter is based on the author's interpretation of information appearing in the official Chinese press and does not necessarily represent the views of any U.S. government organization. A revised version of this chapter appeared in *Pacific Affairs* 56, No. 2 (Summer 1983).

This statement departed significantly from China's single-minded anti-Sovietism of recent years. Zhao was implying that the United States as well as the Soviet Union was dangerous, since it might subvert and invade others to achieve its aims.

In June, Foreign Minister Huang Hua confirmed this trend in an address on disarmament at the United Nations. Huang omitted previous characterizations (found in similar authoritative pronouncements from 1975 to as late as March 1982) of Moscow being "more aggressive and adventurous" than Washington and hence "the most dangerous source of a new world war." Huang also avoided criticizing the Soviet Union by name, thus reverting to a common practice before 1978 whereby both major powers were criticized indirectly.[3]

In moving toward a more balanced stance vis-à-vis the superpowers, China dropped its strident calls for the building of an international united front (implicitly or explicitly including the United States) aimed at opposing Soviet hegemonism. Vague formulas emerged expressing China's willingness to cooperate with all "friendly" or "peace-loving" forces in achieving broad foreign policy goals of political and economic independence. This trend became evident during Premier Zhao's late May 1982 visit in Tokyo.[4]

This balanced rhetorical treatment was mirrored in diplomatic initiatives. The August 17 Sino–U.S. communiqué, on Taiwan, made only indirect references to common strategic goals, in sharp contrast to earlier communiqués that had stressed joint opposition to (Soviet) hegemonism. The 1982 communiqué stated obliquely:

> In order to bring about the healthy development of U.S.–China relations, the two governments reaffirm principles agreed on by the two sides in the Shanghai communique and the joint communique on the establishment of diplomatic relations.[5]

By November, comments by Huang Hua at the time of Brezhnev's funeral implied that Peking placed nearly as much importance on improving Sino–Soviet relations. He expressed, for the first time, that Chinese leadership "attached importance" to the consultations between vice foreign ministers that began in October 1982. He also stated that improved Sino–Soviet relations would be "conducive to peace and stability in Asia and the world as a whole."[6]

General Secretary Hu Yaobang spelled out the import of Peking's new foreign policy line at the 12th Party Congress in September. He depicted threats to China's security and threats to China's sovereignty and independence as of equal concern to the leadership. At the same time, he projected a willingness to be friends with both Washington and Moscow if they demonstrated good intentions toward China with "deeds, not words."[7]

Handling of the Sino–U.S. communiqué and the resumption of talks with Moscow served to underscore China's intent to deal with each country on its own merits and not to allow either to use China as a pawn in some geopolitical game. Chinese official explanations of the Sino–U.S. communiqué stressed the necessity for the U.S. to prove its sincerity through faithful implementation of its provisions. The communiqué focused on bilateral rather than strategic (anti-Soviet) issues, and the day it was signed, a Chinese Foreign Ministry official was in Moscow, presumably discussing plans for reopening talks.

When the talks opened in October, Chinese officials downplayed their importance, insisting that they were "exploratory consultations" rather than formal negotiations like those held in 1979.[8] They also stressed the slim prospects for improving relations without some concession to Chinese demands for a change in Soviet containment policy toward China. Thus, the Chinese insisted on discussing Moscow's support for Vietnam and its occupation of Afghanistan as well as the Soviet troops along the Sino–Soviet and Sino–Mongolian borders.

REASSESSING CHINA'S STANCE

As these modifications in China's foreign policy unfolded, Chinese leaders began to speak in vague terms of an increase in competition between the superpowers. They seemed to be concluding that for the first time in over a decade, Moscow was no longer gaining on Washington in terms of military power and influence. But Peking seemed uncertain about whether this shift would be beneficial to China's interests. This was the import of the comments made in the summer of 1982 by Premier Zhao and Foreign Minister Huang Hua, respectively:

> The current turbulent international situation is marked by ever-increasing tension and turbulence. While old problems remain unsettled, new crises are emerging. Their root cause lies in the rivalry between the two superpowers. On the one hand, one must see that the Soviet Union and the United States are quickening their pace of contention for a strategic superiority in a really aggressive manner. On the other hand the Soviet Union has much more difficulty than it had last year. . .

> One superpower has been pressing forward to expand its sphere of influence. Not wishing to be outdown, the other superpower has exerted its utmost to build up its strength and to try to regain its former position of world supremacy. . .[9]

This assessment is a significant departure from that which underlay Sino–U.S. detente in the 1970s – that the Soviet Union was unrelentingly and successfully expanding its influence, while the U.S. was defending its former spheres of influence with difficulty. Judging that the U.S. would no longer be a threat to China under such circumstances, Peking "leaned to the West." This trend accelerated in the mid-1970s, as Peking feared that U.S. withdrawal from involvement in Indochina would leave a power vacuum in the region that Moscow would move to fill. China quickly moved to improve relations with ASEAN and Japan and encouraged the U.S. to retain its close ties and military presence throughout the region. This trend peaked under Deng Xiaoping's leadership from 1978 to 1980 as the Chinese leadership became alarmed by the invasions of Kampuchea and Afghanistan.

During the late 1970s, Peking also had hopes of large-scale economic cooperation with the West to speed up its development process. Close strategic and economic cooperation with the U.S. and its allies thus appeared necessary, possible, and mutually reinforcing.

There was consistent evidence throughout, however, that some in the Chinese leadership were not convinced of the logic behind this approach, from both the domestic and international angles. In the spring of 1979, in the wake of Sino–U.S. normalization and China's attack on Vietnam, strong evidence of debate appeared in the media.[10] Historical articles used allegories from ancient Chinese history and from early Soviet history to question current policy. (Allegories have long been a favorite mechanism for political debate in China over sensitive issues.)

Some articles seemed to suggest that China could afford neither to "import" modernization nor to pursue a confrontationist foreign policy. Rather it should adopt the approach of Lenin during the days of the New Economic Policy (NEP) and relax restrictions on capitalist practices at home and make amends with enemies abroad. This would gain a "breathing space" for ensuring social stability and building economic and military strength at a slower but steadier pace.[11]

The reasoning put forth in these articles seemed to be mirrored in new policy thrusts in 1979, including an economic readjustment that placed higher priority on slow growth based in agriculture and light industry (which required less foreign involvement than earlier plans); offers to open negotiations with both Moscow and Hanoi to lessen tension and resolve problems; and tighter social and ideological controls. At the same time, a number of officials were rehabilitated who had at one time been criticized for "pro-peace" foreign policy initiatives recommending improved relations with both Soviet and American camps, as well as with all China's neighbors, especially India.[12]

Beginning in mid-1979, as prospects for Sino–American strategic cooperation picked up with Vice President Mondale's visit, and particularly after the Soviet invasion of Afghanistan in December, these voices were stilled. International developments seemed to argue on behalf of the approach associated with Deng. Deng also had strengthened his domestic political position. Even then, however, there continued to be evidence that Deng was acting under constraint in talking abou the potential for long-term Sino–U.S. cooperation.[13]

In late 1980, in the midst of a heated confrontation over the party chairmanship between supporters of Hua Guofeng and Hu Yaobang, a broad-ranging review of Chinese priorities was sparked by economic and social difficulties at home.[14] The review was broadened to cover foreign policy due to events in Poland (which carried important lessons for both domestic and foreign policy), the sale of submarines by the Netherlands to Taiwan, and campaign rhetoric about Taiwan in the U.S. This review continued through 1981.

Allegorical articles again revived the analogy of Lenin's NEP for China's current situation.[15] Again, policy shifts brought economic retrenchment, tighter social controls, and a revival of ideological orthodoxy, as well as a warming trend in low-level Sino–Soviet relations. Moscow's persistent calls for a resumption of talks, and evidence that the Brezhnev succession struggle was well underway, no doubt strengthened the arguments of those who thought China should open the lines of communication.

This lengthy period of leadership evaluation and debate led to the changes in China's foreign policy approach that became evident in 1982. At first, it seemed possible that the critical stance toward the United States was intended solely to increase pressure on U.S. policy makers over the Taiwan issue. This interpretation seemed justified by high-level comment during and after Vice-President Bush's visit to Peking in May that the "major" or even the "only" obstacle to improved Sino–U.S. relations was the Taiwan issue.[16] This implied that a change in Washington's policy toward Taiwan would result in a return to earlier public postures favoring the United States.

But while the August communiqué eased the Sino–U.S. crisis, criticism of the U.S. continued, and it became clear that the change in stance had deeper roots and other targets. The broad range of interrelated changes that took place affected most of China's important relationships. This points to a well-considered decision rather than pressure tactics, since it would be embarrassing if not damaging to China's reputation to rescind them. Moreover, Hu Yaobang and Zhao Ziyang, who are likely to lead China during the next decade, have become personally associated with the construction of the new framework. Thus, the equal stress on China's goals of independence and sovereignty, development, and security is not likely to prove temporary.

At the same time, however, the specifics of China's policy are still evolving and there is much room for flexibility. On balance, implications of the policy seem more practical and less alarming for U.S. interests than the verbiage would suggest. In projecting a close association with third world concerns, greater independence from the United States, and a willingness to deal reasonably with Moscow, Peking has not changed its basic desire for close ties with the West for both security and development purposes. Rather, it is backing away from its single-minded efforts of the late 1970s to build a matrix of relations focused on confronting the Soviet Union.

China hopes thereby to increase its flexibility and expand its options. The main aim is to reduce tensions with Moscow and force caution on Hanoi. But also competition with Moscow in third world and socialist arenas may become more effective. Meanwhile, Washington can be reminded that China has other resources for solving its problems, and thus cannot be taken for granted. The reemergence of problems with the United States over Taiwan since late 1980 served as a catalyst for the modification of China's foreign policy, but the process surfaced earlier and a number of domestic and international factors have shaped the outcome.

RESURGENT SUSPICION OF WASHINGTON

For Peking, Washington's policy toward Taiwan has always been a litmus test of U.S. resolve and ability to orchestrate its foreign policy to achieve the aim of countering Soviet expansion. The Chinese have portrayed recent tensions in U.S.–China relations over Taiwan as a symptom of a larger problem—the reluctance of the Reagan Administration to follow its predecessors in recognizing the necessity for close cooperation (through compromise of often divergent interests) not only with its industrialized allies but with the developing nations. This linkage was made clear in an authoritative *People's Daily (Renmin Ribao)* commentator article pegged to the third anniversary of normalization on January 1, 1982. It asked rhetorically:

> What designs does the United States in fact harbor on China's territory of Taiwan? Should it or should it not be called hegemonism if the United States ignores the principles of international relations, validates the sovereignty of other countries and interferes in their internal affairs in this way?[17]

Tensions in Sino–U.S. relations over Taiwan have precipitated a reassessment of China's foreign policy of the late 1970s—which hinged on a deepening Sino–U.S. relationship—precisely because it raised all the questions that had been swept under the rug during the "honeymoon" of 1979–1980. The ideological conditioning of Chinese leaders and their historical memories of earlier Sino–U.S. conflicts colored this evaluation by heightening their suspicions about U.S. intentions toward China. When Deng Xiaoping in early 1980 spoke of the potential for long-term Sino–U.S. cooperation, he was extending the logic of the immediate post-Afghanistan period into the indefinite future and ignoring questions from his critics about the incompatibilities between the two political–economic systems and about the continuing potential for future conflict with the U.S. and its allies, especially in Asia.

That the wisdom of close ties with the U.S. was being questioned with increasing urgency in 1981 became evident in journalistic and academic assessment in the Chinese media through the year. In early 1981, Peking had appeared optimistic about the prospects for U.S. policy as the media focused on the incoming administration's avowed determination to pick up the Soviet gauntlet and contain Moscow's expansionism. Peking seemed particularly sanguine about the strategic situation in Asia. Low-level commentary regularly gave credit for the favorable regional balance of power to a new pattern of "anti-hegemonist" cooperation among China, the United States, Japan, ASEAN and the countries of Oceania.

But by the end of the year, media assessments were noticeably more downbeat, stressing that Washington's declared hardline approach to the Soviet Union had yet to be worked out in practice. A particularly somber assessment was offered in the January 1982 issue of the *Journal of International Studies* (*Guoji Wenti Yanjiu*), the official journal of the Foreign Ministry's Institute of International Studies. The article noted that economic problems and intraadministration feuding had combined with "policy errors" to produce an American foreign policy devoid of consistency, reliability, and a clear sense of priorities. Among the errors cited were the decisions to lift the grain embargo against the Soviet Union and the U.S. attitude toward the third world, which the article claimed "clearly expresses U.S. super-power essence" and is the "fatal flaw" of Washington's policy.[18]

By mid-1982, Chinese media commentary suggested that the Western alliance was facing its gravest crisis in decades and opined that the Versailles summit had merely papered it over. Peking openly blamed the United States for obstructing third world development and undermining the independence of developing countries.

The more jaundiced view of the United States predominated in 1982, but some earlier official references heralded its coming. Forewarnings included Foreign Minister Huang Hua's banquet toast welcoming Secretary Haig to Peking in June 1981. He made unusually explicit references to areas of disagreement, such as the Middle East and South Africa.[19] Thereafter, the media began to complain about Washington's reluctance to give up "four old friends" (Taiwan, South Korea, South Africa, and Israel), which were depicted as America's anachronistic proxies left over from an earlier era of U.S. imperialism. Thereafter, low-level comment on the United States began to mention a "superpower mentality" or an addiction to "power politics" among influential U.S. figures or groups.

In 1982, the number of official references to negative U.S. intentions toward China began to grow, suggesting that at least some in the Chinese leadership were highlighting divergences between American and Chinese interests. For example, Vice Chairman Li Xiannian, often a spokesman for party conservatives, reminded his audience at a large rally in late January that hegemonism was not to be defined solely in terms of expansion and aggression (as Deng's supporters have attempted to do). He said:

> We firmly oppose hegemonism, which attempts to dominate the world and interfere in the internal affairs of other countries, and also firmly oppose expansionists which invade and occupy the territories of other countries. It is absolutely intolerable that anyone try to encroach on China's national sovereignty, interfere in our internal affairs and obstruct the reunification of our country.[20]

Also, an important article in the April issue of the *Journal of International Studies*, reprinted in *People' Daily*, reviewed the Korean War period and implied that the United States was still pursuing a containment policy toward China. The article first declared that sections of the Taiwan Relations Act were intended to justify U.S. obstruction of Taiwan's reunification with the motherland and even U.S. armed intervention for that purpose. It claimed that "some people are trying to control Taiwan forever and use Taiwan to pin down the PRC."[21]

Increased U.S. support for Seoul under the Reagan Administration also heightened suspicion about U.S. intentions in Asia. This played a part in spurring Peking to repair its relations with Pyongyang, which had been strained by Sino–U.S. normalization (among other things). In April, Deng Xiaoping and Hu Yaobang made a secret trip to Pyongyang to discuss foreign affairs and aid and trade, as well as domestic issues. In June, at the time of the Korean War anniversary, and after Secretary Weinberger's visit to Seoul, Geng Biao made the first visit to Korea in decades by a Chinese Minister of Defense and blamed tensions in the peninsula on U.S. "hegemonist" designs in the region.[22] Kim Il-sung's visit to Peking in September capped this trend of improving relations, although Chinese leaders were careful to stress that Sino–North Korean cooperation was aimed at building regional peace and stability.

During 1982, the Chinese leadership was also forced to consider two other sensitive problems—the matter of how and when to reassert sovereignty over Hong Kong and (what Peking viewed as) the reemergence of militaristic sentiment in Japan.[23] These matters had to be addressed in preparing for September visits by Prime Minister Thatcher and Prime Minister Suzuki, both in celebration of a decade of normal relations. The Taiwan problem and the Falklands War (in which China had sided publicly with Argentina) left neither British nor Chinese officials in the mood for compromising. In the prevailing atmosphere, Chinese officials also felt compelled to speak out early and loud against the revisions in Japanese textbooks downplaying Japanese aggression in World War II, in part due to a desire to make Taipei officials appear remiss as nationalists. In the course of 1982, Peking also ceased to applaud U.S.–Japanese security arrangements, including U.S. encouragement of increased Japanese defense spending.

The resurgence of suspicion with regard to Western and Japanese long-term intentions toward China has been accompanied by discussion in China of problems in maintaining cultural and economic independence under new liberal trade policies. Delays in promulgating regulations for foreign investment in China, as well as references to Western "bourgeois ideology" as the alleged cause of widespread economic corruption in China, have provided indirect evidence of continuing pressure on the Chinese leadership to tread cautiously in this area. In discussing the anticorruption campaign, for example, an article in the *Intellectuals' Daily* in June 1982 said that "imperialism and hegemonism are always trying every possible means to penetrate, sabotage and subvert our country's politics, economy and culture."[24] And a *Red Flag (Hongqi)* article in mid-May on corruption, written by its editorial board, warned that "class enemies at home and abroad are trying by every possible means to sabotage our socialist modernization and disrupt our communist movement."[25] These undercurrents, the strikingly defensive tone of official statements with regard to the necessity for continuing a long-term "open door" policy, and complaints in late 1982 about U.S. reluctance to transfer advanced technology to China strongly suggested that some conservative party elements in China have tried to blame China's economic and social problems on Western efforts to sabotage Chinese development efforts.

In this political context, the party's reform wing led by Deng Xiaoping tried to gather evidence of the West's willingness to deal with China as an equal and a friend and to eschew efforts to subvert China's political and economic system, as well as evidence of the development gains to be made through the open door policy. Circumstantial evidence suggested the leadership had trouble on both scores, particularly with regard to the United States. The Chinese media persistently complained that Washington refused to deal with Peking on a basis of mutuality. They drew implicit comparisons between current trends and the split between Moscow and Peking in the early 1960s when the Soviet Union allegedly attempted to make China its satellite through political and economic pressure.

EMERGING REALISM IN DEALING WITH THE SOVIET UNION

There is no doubt that the Soviet Union continues to be viewed as the most serious potential threat to Chinese security, despite the reemergence of problems with the West. Modifications in Chinese policy were undertaken with that fact foremost in mind. The major combined forces exercise carried out in northern China in late September 1981 underscored this point. An editorial in the army newspaper commenting on the exercise as it was underway warned that "Soviet hegemonism is speeding up its global strategic deployments and seriously threatening world peace and China's security."[26] And Chinese officials highlighted the direct Soviet threat to China even as bilateral talks resumed. Moreover, commentary through 1982 about the complexity and instability of the international situation indicated that Peking was not sanguine about the long-term prospects for peace.

Peking apparently concluded, nevertheless, that something might be gained from talking to Moscow, something more than a temporary avoidance of hostilities. Earlier offers to talk, in 1969 after border clashes and in 1979 after Sino–Vietnamese hostilities, seemed intended by the Chinese primarily to reduce immediate tensions. This time, the Chinese seemed hopeful that Soviet difficulties with the United States, in Eastern Europe and in Afghanistan and Indochina, combined with a need for calm during the succession to Brezhnev, might induce Soviet concessions to Chinese interests in Asia for the first time.

To encourage Soviet willingness to change, Peking signaled a strong interest in making progress possible by hinting for the first time that a concrete Soviet "gesture" on only one or two of China's demands might be sufficient for improvement in bilateral relations. The atmosphere for the talks was more carefully prepared than in 1979 through diminishing polemics, the overall distancing from the United States, and resumption of a wider range of nonpolitical contacts. Meanwhile, media commentaries prepared the domestic audience for a gradual improvement of Sino–Soviet relations.[27]

This willingness to probe Moscow contrasted sharply with the approach during the height of Peking's calls for a global coalition against the Soviet Union. Chinese officials then warned that any willingness to negotiate would be seen as a sign of weakness by Moscow and thus would incite more "bullying." They railed against Western policies of detente, calling them appeasement.

In 1980, following the Soviet invasion of Afghanistan, Chinese commentary stressed the immediate dangers posed by what they called a Soviet "southern strategy" aimed at cutting strategic links between the U.S. and its allies through expansion into Southeast Asia. By 1981, Chinese officials seemed on the defensive in arguing that this development was inexorable. Vice Foreign Minister Zhang Wenjin at the United Nations in September, for example, said that:

> It would be contrary to the international scene to consider that the Soviet Union is on the defensive and that its deep predicament is forcing it to consider a retreat. The facts in the past year have shown that the Soviet Union has not given up its bid for world hegemony and the corollary strategy of a southward drive remains unchanged.[28]

A *People's Daily* commentator article on March 21, 1982 turned out to be the swan song for this analysis. It also was the last authoritative reference to Peking's assessment for nearly a decade that the Soviet Union was "the most dangerous source of war." Trends in Chinese commentary after March implied, in contrast, that some acts of Soviet expansion may have been due to opportunism rather than a global strategy. Thus, presumably, negotiations might play a role in reversing them.[29]

This shift in assessment of the Soviet threat and how to deal with it has been buttressed by an evolution in the ideological treatment of the Soviet Union. In the mid-1970s, Mao posited that the Soviet system was "revisionist," not socialist. This corrupted Soviet foreign policy, making it "social imperialist." This analysis underlay Chinese expectations that the Soviet state system, with its militarized economy, would inevitably drive its expansionism.

By late 1979, references to revisionism had been dropped as an embarrassment to Deng Xiaoping's wing of the party, which had been purged by Mao for following the Soviet Union in "taking the capitalist road." This left unanswered the question of whether the Soviet Union was socialist. On the eve of the People's Republic of China's 30th anniversary, Vice Chairman Ye Jianying in September 1979 defined Soviet social–imperialism as repression at home and expansion abroad, without attempting a Marxist explanation of the linkage.[30]

A theoretical explanation was attempted in a *Red Flag* commentator article in mid-February 1980, which listed standard Chinese grievances against Soviet acts of aggression from the invasion of Czechoslovakia to the invasion of Afghanistan.[31] It traced the roots of Soviet global hegemonism to historical Russian chauvinism. Citing Lenin, the article depicted the expansion of the czarist empire as the gradual Russian annexation of small and weak neighboring political entities and the suppression of their non-Russian peoples. The deep-seated influence of czarism on the post-Stalin leadership, according to the article, had led to pursuit of "political reaction" at home and "reactionary nationalism" abroad that "placed the interests of the ruling group above those of the people of other nations." Thus, the Soviet Union had degenerated into a social-imperialist country bent on subjugating other sovereign states and inciting others such as the Vietnamese to pursue similar policies. While such an analysis certainly implied that rapid change in Soviet behavior was not likely, it did leave open the possibility that Soviet leaders could overcome the weight of history and adopt new policies of "progressive nationalism."

In 1982, the Chinese leadership had moved further in this direction by showing a willingness to restore ties with communist parties, such as the French and the Dutch, who value the Soviet model and Soviet policies. They also admitted to the past value to China of close ties with the Soviet Union and dropped the term "social–imperialism." The sole remaining issue of contention between the two parties was Moscow's "hegemonist" attitude, toward both other states and other parties. In mid-October 1982, Hu Yaobang announced that the Sino–Soviet talks would continue, and he added that the Chinese Communist Party (CCP) would establish contacts with "any party willing to be friendly with us."[32] The message to the Communist Party of the Soviet Union (CPSU) seemed to be that the CCP was willing to make amends if it were agreed to set differences aside.

In a striking conciliatory move in November, Foreign Minister Huang Hua refrained from assigning blame for the Sino–Soviet split to the Soviet leadership. In remarks that strongly implied that the Soviet Union should be considered socialist, Huang called Brezhnev an "outstanding statesman" and expressed Chinese "appreciation" for his gestures toward China and "hope" that his successor would move further in this direction.[33] Chinese interest in improving state relations and resuming party contacts seemed stronger than at any time since the 1960s.

DOMESTIC POLITICAL PRESSURES

From late 1980 until early 1982, as the Chinese leadership engaged in a wide-ranging review of domestic and foreign policy problems, Deng Xiaoping was on the defensive. Debate over foreign policy was spurred on by a steady stream of mixed signals from the Reagan Administration with regard to China policy. The year 1981 was marked by retrenchment in most fields, as Deng's reform wing of the party regrouped by defending its basic program but making concessions in accordance with the views of more conservative leaders.

During this period of review, the Chinese leadership made few authoritative statemens with regard to foreign policy. Given the primacy of domestic issues as Deng worked to strengthen the positions of Hu Yaobang and Zhao Ziyang, international affairs was of secondary concern except for highly politicized problem areas (most notably, the reemergence of tension with the United States over Taiwan).

Indirect evidence that Deng's critics saw his "peaceful" Taiwan policy as his achilles heel came in historical articles that revealed that Deng was being criticized for writing off China's hopes of regaining Taiwan in order to improve strategic and economic relations with the United States. For example, in late December 1980, an article in the academic journal *Studies in Chinese History (Zhongguoshi Yanjiu)* described in detail the careful preparation over a 10-year period by the Kang Xi emperor for the takeover of Taiwan.[34] It depicted heated debate in the Qing (Ching) court over issues with current resonance: whether the recovery of Taiwan was essential for ensuring the security and prosperity of southeastern China; whether regaining Taiwan was militarily feasible and (once recovered) defensible against foreign powers or should perhaps be abandoned in recognition of Qing weakness; and what level of autonomy should be allowed the island's rulers after assertion of Qing control.

An even more striking article appeared in People's Daily on January 8, 1981.[35] It pointedly insisted that "we should... learn from" an earlier proposal of the late Premier Zhou En-lai (not heeded at the time) to attack Chiang Kai-shek in the wake of his annihilation of Shanghai's Communists in 1927. The article seemed to be implying that today China should use the threat of force against Taipei rather than "hesitate, intending to settle problems by relaxation and preparation for long-term struggle." The latter approach would only result in "Chiang's political power... becoming more consolidated and his relationship with the imperialist powers gradually growing closer." The article also seemed to argue by analogy that China could feel free to deal more harshly with Taipei, because the United States would not intervene.

Sensitivity over the Taiwan issue was reflected throughout 1981 in the appearance of other historical articles dealing with themes of patriotism and national betrayal in China's historical relations with the West and Russia. While some articles seemed genuinely part of an ongoing reassessment of history, others had overtones that seemed to be questioning Deng's policies through the use of allegory. An article in the November 23 issue of the *Guangming Ribao (Intellectuals' Daily)*, for example, severely criticized a group of late nineteenth century officials for pursuing appeasement policies that pursued negotiation of formal but unequal agreements with the West in hopes of obtaining technology and military equipment, thus allowing Western encroachment on China's sovereignty.[36] It praised another group of officials who combined negotiation with "preparations for war to check the enemy's aggression on the basis of military strength."

In the face of persistent questioning, Deng's primary concern in foreign policy in 1981 was to defend the rationale of his basic approach premised on close ties with the United States, and to quickly resolve problems. But his repeated failures to obtain a clear commitment from the new U.S. administration to the "one-China" principle that underlay normalization put him and his associates in an increasingly difficult position.

A remarkable article in the October 1 (National Day) issue of the academic journal *World Knowledge (Shijie Zhishi)* seemed to be a transparent effort to defend Deng.[37] It claimed that the formation of a consistently patriotic policy since 1949, after a century of national betrayal, was the product of a series of power struggles between pro- and anti-independence forces. The article explicitly praised the policies of Deng Xiaoping as well as of Mao Zedong and Zhou Enlai in the face of slander (attributed to foreign sources) of China's international stance as unprincipled and unpatriotic. The article defensively argued that in dealing with continuing differences with Washington, "the Chinese people had firmly and repeatedly expressed their own principled stand and carried out necessary struggle to safeguard the long-term objective of national unification." It declared that "we will never change our objectives because of the establishment of diplomatic relations with the United States."

In 1981, Hu Yaobang also found it necessary to prove his own determination to protect China's national interests. Speaking at a July 1 rally on the party's 60th anniversary, Hu assured the assembly that the leadership would not tolerate "any servility in thought or deed" in the face of "hegemonist threats of force" and in relations with "all stronger and richer countries." He juxtaposed to this pledge a reiteration of the vow to reunify Taiwan with the mainland, making the reference quite pointed. A similar tone was present in Hu's speech on October 10 at a rally commemorating the 70th anniversary of the 1911 revolution.[38]

Shortly thereafter, Peking's proposal to begin talks on peaceful reunification with Taipei and its demands that the United States agree to limit arms sales to Taiwan marked a serious major effort by Deng and his colleagues to achieve a breakthrough toward resolution of problems with Washington. They hoped to remove the Taiwan issue from the immediate political agenda and thus help to regain their domestic political momentum.

A turning point came in early 1982. Deng Xiaoping then was having difficulty pushing through personnel changes in the central party and government organs that would strengthen his base of support during preparations for the 12th CCP Congress. At that point, in January, the Reagan Administration announced that the United States would not sell Taiwan the more advanced fighter it had requested, but would renew arrangements for coproduction of the F5-E. This offered the possibility, but not the certainty, of avoiding a downgrading of relations. In March, the two sides finessed the handling of a U.S. sale of military spare parts to Taiwan so as to continue negotiations on the larger issue. During this period, however, the Chinese leadership seemed to conclude that even if the immediate crisis could be eased, the Taiwan issue and other problems with the U.S. were likely to continue to demand attention rather than recede to the background. Pressure on the leadership to address the recovery of Hong Kong and to redress the problems attending growing Western involvement in China also pointed in this direction.

In this situation, the Dengists gave ground to their critics by agreeing to project a more neutral stance vis-à-vis the superpower camps and resume talks with Moscow, in part to prove that they were neither provoking Soviet hostility nor being taken advantage of by the West. In initiating these moves, however, Deng's group hoped to carefully manage the shift so that it had minimum negative impact on relations with the West, but served several immediate tactical purposes: to make clear to Washington that problems needed to be resolved and to undermine Hanoi's confidence in its support from Moscow as the season for fighting in Indochina approached, thus strengthening the chances for survival of the new coalition government in Kampuchea. Perhaps most important, lowering tension with China's neighbors was essential, especially given Sino–U.S. problems, in order for Deng to press on with his urgent domestic agenda: a major restructuring of central and local organs, including the military command, and a major turnover in party personnel. These potentially destabilizing reforms would be hard to justify in an atmosphere of regional crisis or uncertainty.

Through 1982, Deng and his colleagues appeared to be balancing their concerns fairly well. Vice President Bush's visit in May and letters to Chinese leaders from President Reagan in the summer served as evidence that the United States would work toward a resolution of the crisis by aligning its Taiwan policy with U.S. commitments to China. By taking strong stands on the Japanese textbook controversy and the sovereignty of Hong Kong, Deng's men moved both matters from the political agenda, so that they could be handled with flexibility through diplomatic channels.

By the time of the preparatory meetings for the CCP's 7th plenum in late July, Deng had regained enough political momentum to set a 12th Congress agenda that would greatly strengthen Hu Yaobang's position. He also obtained a mandate to compromise once more with Washington over Taiwan. The timing of events suggests that Deng used the constitutional requirement that a party congress be held every 5 years to force through these decisions in time to beat the August deadline for President

Reagan's notification of the U.S. Congress on the F5-E decision with regard to Taiwan. He could then portray the communiqué as a victory, going into the Congress.

Once the communiqué and the 12th Party Congress were safely behind him, Deng resumed Sino–Soviet talks in early October 1982, without serious risk to Sino–U.S. relations. Since the talks were held at the time of high-level visits from Japan, Europe, Pakistan, and Thailand, moreover, the Dengists could use them to reassure the West of China's continuing interest in close ties and to allay fears of a major shift in China's strategic alignment. Thus, the Sino–Soviet thaw served as a reminder to Washington of China's importance, but not as an effort to blackmail the United States into the communiqué.

In the course of these events, Hu Yaobang gained sudden and surprising influence in foreign affairs. The restructuring of central organs resulted in considerable turnover among the key actors (from the Secretariat, the Politburo, the Military Commission, the International Liaison Department, and the State Council) who are involved in making foreign policy. Officials who had worked with Hu Yaobang in youth organizations in the 1950s and 1960s were placed in key posts. Several, including the new Foreign Minister, Wu Xueqian, had worked in international liaison affairs for both the party and its youth organization. Their experience may have predisposed them to recommend to Hu a foreign policy centered on enhancing China's influence within the socialist and third worlds. Such a policy is consistent with Hu's reassertion at the 12th Congress of China's orthodox commitment to promoting communist morality and eventually achieving communist goals at home and abroad.

Whether these concerns of Hu's group will prove compatible with a pragmatic development of good relations with all countries for the sake of diversifying economic ties, an approach more closely identified with Premier Zhao Ziyang, remains to be seen. Perhaps tensions caused by the intrinsically different bureaucratic interests of party and government leaders, evident in the differing approaches to foreign affairs by Mao and Zhou, are inevitable. Hu Yaobang must establish himself as an orthodox Marxist leader and would naturally concern himself more with China's image as a socialist nation. Zhao's future rests heavily on economic progress and the successful management of foreign (primarily Western) economic ties toward this end.

Difficulties in reconciling the various aims of China's new foreign policy, and perhaps continuing dispute over those aims, may account for ambiguities in China's treatment of the two "hegemonist superpowers" through late 1982. Zhao seemed consistently more willing than Hu to distinguish between the superpowers on the basis of the threat they posed to Chinese interests. For example, Hu Yaobang, in October, flatly refused requests to expound on any such differences.[39] Zhao, in contrast, took pains to reassure Prime Minister Prem of Thailand in November that China continued to have common strategic interests with the United States despite its isolated "acts of hegemonism" and bilateral problems. He saw Soviet threats to Chinese security, however, as symptomatic of a global hegemonistic policy. Zhao failed to express any interest in better relations with Moscow, but called on Washington to "make new efforts" to improve Sino–U.S. relations.[40]

CONCLUSION

Chinese uncertainty about international trends, particularly U.S. policy toward China, has sparked continuing debate in the leadership over foreign policy in recent years. China's new stress on "independence" represents a consensus position that all can agree on. Specific foreign policy initiatives in 1982 appeared intended to stabilize or improve all China's foreign relations, pending developments that could resolve the debate. These motives were reflected in comments by officials in 1982 that they were "hoping for the best while preparing for the worst." While Peking continued to negotiate with Washington on the Taiwan issue, it worked to cement good relations with other industrial nations. It reaffirmed ties with old friends such as Pakistan, Romania, and North Korea and worked to improve relations with close neighbors including Thailand and India as well as Japan. During reciprocal visits, Chinese officials promised China's cooperation "no matter how stormy and unstable the international situation may be" and "no matter what happens."

China's efforts to expand its options and secure its interests also involved expanding ties with socialist and progressive parties and improving relations, beginning with technical and cultural contacts, with the Soviet bloc. The thaw with Moscow was fueled by a desire to encourage Soviet concessions during a succession period and to explore Soviet bloc sources of input for China's development process.

The moves toward a more pragmatic and flexible foreign policy give higher priority to protecting and expanding China's foreign economic ties. This shift in priorities was made explicit in a lengthy authoritative article in the mid-April 1982 *Red Flag*, which said:

> For a long time, we upheld the argument that economics must be subordinated to politics. This argument is partly right but it fails to take every aspect of the problem into account. One of the basic viewpoints of Marxism–Leninism is that economics and politics influence and act on one another but in the final analysis, politics is decided by economics. This is also true in handling the issue of foreign relations. If foreign economic relations are developed in a satisfactory manner, it will facilitate the continuous development of political diplomatic relations. On the contrary,

if we fail to open up the prospects for developing foreign economic relations, the development of political diplomatic relations with foreign countries will inevitably be obstructed and will never be able to be vigorous.[41]

The type of pragmatic nationalism and down-to-earth Marxism preached by Hu Yaobang places the highest priority on economic development. And thus to ensure the Western input that is essential for sustained economic progress, Hu is likely to handle carefully any conflicts of interest. Yet important constraints on his ability to be flexible derive from his need to build legitimacy as a leader. He must not only protect China's security from direct military threat but also safeguard its integrity as a national unit and as a socialist system from more indirect undermining.

As long as Taiwan and Hong Kong remain outside the exercise of Chinese sovereignty, they will be the focal point for tension that stems from the larger unanswered questions about China's relations with the West: to what extent Chinese, American, and Japanese interests in Asia coincide; to what extent can the Western and Chinese economies be made compatible; would Sino–U.S. military cooperation benefit both parties or alarm allies and friends of both and provoke Moscow; will Chinese links with the West lead to dependence that will undermine the Chinese political system; can China's outstanding territorial claims be resolved to its satisfaction by peaceful means?

In the euphoria of normalization, and under the cloud of Soviet threats to Poland and Afghanistan, both China and the U.S. avoided addressing these questions. Both sides raised unrealistic expectations about the degree of mutuality of interests. Certain aspects of President Reagan's policies contributed to the return of these hard issues to the limelight in Peking.

Ironically, as policies in the United States moved closer toward the single-minded anti-Soviet confrontationism that China had been urging for so long, Peking changed its mind. Events in Poland, the Middle East, and Central and South America found China more closely aligned with some European positions, or even with Moscow, than with the U.S. stance. Media commentary suggested that the Chinese became more sympathetic than ever before to the foreign policy approaches of France and West Germany, which value detente with the Soviet Union. Peking has also expressed enthusiasm for Yugoslavia's posture of nonalignment, which allows for broad-ranging ties with East and West. While China's newfound neutrality is not a return to the more ideological and isolationist policies of Mao Zedong, it is likely to prove an important and lasting retreat from the "lean to the West" that emerged briefly under Deng Xiaoping.

The most important lesson of the last few years for the new Chinese leadership emerging from under the tutelage of Deng Xiaoping appears to be that assumptions of inevitability are handicaps to policy. Trends in Sino–American and Sino–Soviet relations, as well as in China–Taiwan relations, are not set in the cement of history. The Reagan Administration brought home to the Chinese the fact that trends set in place by previous administrations could be reversed by a succeeding one. Faced with the upcoming successions to Deng and Chiang Ching-kuo (in Taiwan), as well as Brezhnev, Peking in 1982 was learning to expect the unexpected.

NOTES

1. *Xinhua* in English (hereafter cited as *Xinhua*), April 18, 1982, in Foreign Broadcast Information Service *PRC Daily Report* (hereafter cited as FBIS) April 20, 1982, I2. For the text of Zhao's speech on October 22, 1981 at the International Conference for Cooperation and Development at Cancun, see *Renmin Ribao*, October 24, 1981, p. 1. No special importance was attached by Peking to Zhao's comments at the time. Later, *Beijing Review* No. 18, 1982, highlighted Zhao's comments on foreign policy in a special box, but without making clear that they were new. Much later, *Ban Yue Tan* No. 12, June 25, 1982, in FBIS July 2, 1982, A3, published an article by its editorial department positing that the "three basic points" put forth by Zhao on April 18 were "based on" Mao Zedong's "three worlds" theory and would "remain the guiding ideology for our country's foreign policy in the future." (Thus the fiction has been maintained that Chinese foreign policy has been consistent since the mid-1970s.) China's new focus was explained in *Hongqi* No. 13, 1982, and the trend accelerated thereafter.

2. *Xinhua* April 19, 1982, in FBIS April 20, 1982, I6. A short commentary in *Renmin Ribao*, July 9, 1982, in FBIS July 12, 1982, I1, cited these remarks, underscoring their importance. *Xinhua* May 31, 1982, in FBIS June 1, 1982, D3, also depicted Zhao's talks that day in Tokyo with Prime Minister Suzuki as giving nearly equal blame to the superpowers for current international tension.

3. *Xinhua* June 11, 1982, in FBIS June 14, 1982, A1-4. Huang Hua was addressing the second U.N. Special Session on Disarmament. Shortly thereafter, in Bonn, Huang spoke of a "drastic" change in the international situation since October 1981, without elaboration.

4. See note 2. Only 6 months earlier, at ministerial talks in Tokyo, Huang Hua had called boldly for "China, Japan, West European countries, and the U.S., as well as all countries that are opposed to Soviet aggression and expansion" to "keep vigilance, strengthen consultations and coordinate actions" to support the Afghanistan and Kampuchean peoples against aggression. Reporting on Zhao's trip indicated no such references. See *Xinhua*, December 16, 1981, in FBIS December 16, 1981, D4.

5. *Xinhua* August 17, 1982, in FBIS August 17, 1982, B1.

6. *Xinhua* November 18, 1982, in FBIS November 18, 1982, C1, and *Xinhua* November 14, 1982, in FBIS November 15, 1982, C2. Huang's last comment, which signaled that Chinese foreign policy is not necessarily aimed against Moscow, but could encompass friendly Sino-Soviet relations, first appeared in China's annual greeting message to the Soviet on the anniversary of the Bolshevik Revolution in 1982. See the Peking broadcast in Russian, November 7, 1982, in FBIS November 8, 1982, C1.

7. *Xinhua* September 7, 1982, in FBIS September 8, 1982, K18-21.

8. Until near the end of the first round of talks, the Chinese media did not report that they were underway. Rather, Chinese officials gave their views of the talks to foreign reporters. *Xinhua* October 17, 1982, in FBIS October 18, 1982, C1, gave the first official report on the talks in citing Hu Yaobang's statement to French journalists that the "consultations" would be continued in Moscow. In late 1979, "negotiations" to improve relations were conducted by vice foreign ministers, but came to naught, as Peking demanded that Soviet policies in Mongolia and Indochina be included on the agenda. Moscow refused to discuss issues involving third parties. In January 1980, Peking stated that it was "at present inadvisable" to hold the second round of talks due to the Soviet invasion of Afghanistan. The resumption

of talks in a more informal format has broken the stalemate without either side having to give away openly on its principles.

9. See notes 2 and 3.

10. Editorials in late March seemed especially defensive about the attack on Vietnam, with one arguing that "it would not do if we had not fought." See *Renmin Ribao* and *Jiefangjun Bao*, March 26, 1979, in FBIS March 27, 1979.

11. For an example, see the article by Wu Qiawen in *Zhexue Yanjiu* 1979 No. 4, pp. 21–26. References to the 1918 Treaty of Brest–Litovsk has almost become a slogan for those advocating this point of view. References to "seeking common points while reserving differences" and discussion of the dialectical theory of the unity of opposites have resurrected debates from the mid-1960s that touched on how to handle the Sino–Soviet split.

12. An important article in *Gongren Ribao*, April 5, 1979, in FBIS May 4, 1979, L13, served as a posthumous rehabilitation of former CCP leader Wang Jiaxiang. The article appeared 2 days after China proposed to open talks with the Soviet Union on the normalization of relations and served as a justification for the move. It defended Wang's proposals in the early 1960s and again in the early 1970s to reduce tensions with all China's neighbors for the sake of economic recovery. The decision to exonerate Wang was probably made at the CCP's 3rd plenum in December 1978. At that time, two other officials who had allegedly supported Wang's views in the 1960s, Chen Yun and Yang Xianzhen, were returned to influential posts.

13. At the time of Defense Secretary Brown's visit to China, a most unusual "mistake" was made by *Xinhua* in reporting a banquet toast by Deng Xiaoping in which he called on all nations to forge an "alliance" against Moscow. *Xinhua* later "corrected" this to read "to unite." See *Xinhua* January 8, and 10, 1980, in FBIS January 8 and 11, 1979, B1. On April 11, 1980, Deng met with AP correspondent John Roderick and made a rare reference to "long-term strategic" relations between China and the United States. The fact that other Chinese leaders have been much more reluctant to speak of anti-Soviet cooperation, especially on a lasting basis, strongly suggests that Deng has been in the minority on this issue all along.

14. A party work conference that convened in December 1980 was originally intended to discuss political matters, but the agenda was changed according to Hu Yaobang, speaking at a New Year's reception. See *Xinhua* January 1, 1981, in FBIS January 2, 1981, A5. An editorial in *Renmin Ribao*, January 19, 1981, in FBIS January 19, 1981, L4, defensively asserted that the policies (including foreign policy and foreign economic policy) of the 3rd plenum would never change. Since one of the most notable policies endorsed at the plenum was Sino–U.S. normalization, this editorial was rare direct evidence of debate over foreign policy and probably over Sino–U.S. relations.

15. See Lu Ding, "Advance and Retreat," *Hongqi* No. 1, 1981, in Joint Publications Research Service (JPRS) *China Report (Red Flag)* 77587, March 13, 1981, p. 44.

16. See Vice Premier Wan Li's comments upon Bush's departure, *Xinhua* May 8, 1982, in FBIS May 10, 1982, B2, and Premier Zhao's interview with *Xinhua*, May 14, 1982, in FBIS May 17, 1982, A1.

17. *Renmin Ribao*, December 31, 1981, in FBIS December 31, 1981, B1.

18. *Guoji Wenti Yanjiu* No. 1, 1982. A review of Reagan's foreign policy later in the year, by an analyst at the U.S. Research Center of the Academy of Social Sciences, took these arguments much further. See *Renmin Ribao*, July 31, 1982, in FBIS August 2, 1982, B1–6. The article claimed that the Administration's foreign policy was "divorced from complex reality," since it regarded the solution of all international problems as "revolving around two poles. . . . It simply ascribes all international turmoil to Soviet meddling and never shows proper concern and understanding for the interests of its allies and the complex conditions in the Third World." It concluded that "The nature of a superpower determines that the contradictions in the Reagan Administration's foreign policy are inevitable."

19. *Xinhua*, June 14, 1981, in FBIS June 15, 1981, B2.

20. *Xinhua* January 24, 1982, in FBIS January 25, 1982, K2.

"Independent" Chinese Foreign Policy / 83

21. *Guoji Wenti Yanjiu* No. 2, in *Renmin Ribao*, April 6, 1982, in FBIS April 6, 1982, B1. The article was authored by a "contributing commentator," probably a high-level official.

22. *Xinhua* June 15, 1982, in FBIS June 16, 1982, D1. The April visit was not publicized until Kim's September visit to Peking. See *Xinhua* September 17, 1982, in FBIS September 20, 1982, D1.

23. In the previous year or so, the 15-year period used for leases and investment agreements began to force Hong Kong businessmen to think about the prospects for Hong Kong beyond 1997. At that time, the Qing government's agreement to lease the New Territories to the British would lapse. Although the People's Republic has never recognized the legality of any of the agreements (including those granting British control over Hong Kong island and Kowloon), the 1997 date still forces the Chinese and the British to work out some sort of a settlement of the sovereignty issue.

The issue of Japanese "militarism" resurfaced, after lying dormant since the early 1970s, when the Ministry of Education revised official textbooks to call the Japanese invasion of China an "advance" into China, among other changes. The matter became a serious domestic political issue in Japan, as most countries in Asia protested vehemently. In China, local meetings were held in areas that had come under Japanese control to review the history of Japanese atrocities and protest the textbook revision, along with other evidence of sentiment in favor of former Japanese imperialism.

24. *Guangming Ribao*, June 2, 1982, in FBIS June 14, 1982, K3.

25. *Hongqi* No. 10, 1982, in JPRS *China Report (Red Flag)* 81291, July 16, 1982, p. 1.

26. *Jiefangjun Bao*, September 17, 1981, in FBIS September 28, 1981, K2.

27. Hu Yaobang, in his work report to the 12th Party Congress, hinted at a process of improving relations that seemed to involve gradual and reciprocal "steps" rather than Soviet capitulation to China's demands. Once talks began, Chinese officials more clearly indicated to foreign reporters and visitors that a concrete Soviet gesture with regard to one or two of China's demands would be sufficient to keep the process going. For example, Vice Foreign Minister Qian Qichen, in an interview with Austrian journalists in Peking on December 6 spoke of the talks with his counterpart in October, and said "It is hardly possible for the Soviet Union to solve all these problems at once, but even if there were changes with respect to one or another item this would lead to an improvement of relations." See *Die Presse*, December 7-8, 1982, in FBIS December 7, 1982, C1.

Thus, Peking returned to the approach first tried in 1979 with Moscow, which was analogous to the approach used whenever China has wanted to make genuine progress in a relationship. That approach involves setting aside the most contentious issues while seeking improvement in nonpolitical relations, with the hope that after the atmosphere cleared and trust was rebuilt, the hard issues could be resolved peacefully. This return to "people's diplomacy" was stepped up in June 1982, when an official Chinese trade promotion delegation visited Moscow and Soviet athletes attended an international meet in Peking. Hu Yaobang at the 12th Congress stressed that China would strive to develop friendship between the two peoples no matter what the status of state relations happened to be.

Domestic readers were alerted to this new approach when *Renmin Ribao* serialized an article stressing China's desire for peaceful relations with all its neighbors, according to Zhou Enlai's foreign policy of making friends through people's diplomacy. It quoted Zhou as saying that "For countries which hold views different from ours, despite the fact that they might not understand us for the time being, certain contacts were nonetheless necessary to enhance communications and promote mutual understanding." See *Renmin Ribao*, July 10-15, 1982, in FBIS July 14-17, 1982.

28. *Xinhua* September 24, 1981, in FBIS September 24, 1981, A2.

29. *Renmin Ribao*, March 21, 1982, in FBIS March 22, 1982, F1. Comparison of the March commentator article with one that appeared in *Renmin Ribao*, December 1, 1982, in FBIS December 1, 1982, A4, makes the shift apparent. The first article used the Soviet occupation of Afghanistan as proof that the Soviet Union is "the most dangerous source of war,"

because its ambition for world hegemony would cause it to "expand everywhere." Moreover, its burden in Afghanistan "cannot be termed heavy" and will not force it to pull out, because its presence there was "an important cardinal link in Soviet global strategic plans." The December article was much less alarmist and less pessimistic. While it referred to the desire of "Soviet hegemonism" to "thrust southward" and its "reluctance" to pull out of Afghanistan, the article flatly stated that "Afghanistan has become a heavy burden on the Soviet Union." Thus, "the possibility still exists for a political solution." (The March article called talk of a "so-called" political solution merely a "Soviet conspiracy.")

30. *Xinhua* September 29, 1979, in FBIS October 1, 1979, L34.

31. *Hongqi* No. 4, 1980, in JPRS *China Report (Red Flag)* 75525, April 18, 1980, p. 27.

32. See note 8. On the eve of his visit to Tokyo in May 1982, Premier Zhao had referred to Soviet hegemonism in such a way that it made clear it was a "difference in principle" (that is, a matter of ideology) as well as a problem in state relations. See note 16. But Hu Yaobang avoided taking this strong stance in his 12th Party Congress report, even though he did speak of past suffering by the CCP caused by the "attempt of a self-elevated paternal party to keep us under control." See note 7.

33. *Xinhua* November 14, 1982, in FBIS November 15, 1982, C2. These statements implied that Brezhnev's policies had been in the interests of the Soviet people, when barely 6 months earlier, Peking viewed him as a "fascist." Moreover, they implied that his (wrong) actions toward China in the past might have been caused in part by Chinese errors. Thus, the comments suggested that Chinese leaders were willing to consider the Soviet system and Soviet policies as falling within a broad definition of "socialist."

34. *Zhongguoshi Yanjiu* No. 4, 1980, in JPRS *China Report (Political, Sociological and Military Affairs)* 78012, May 6, 1981.

35. *Renmin Ribao*, January 8, 1981.

36. *Guangming Ribao*, November 23, 1981.

37. *Shijie Zhishi* No. 19, 1981, in FBIS November 19, 1981, A1.

38. *Xinhua* July 1, 1981, in FBIS July 1, 1981, K52; and *Xinhua*, October 9, 1981, in FBIS October 9, 1981, K3.

39. See note 8.

40. *Xinhua*, November 19, 1982, in FBIS November 19, 1982, E1. Other evidence of continuing ambiguity regarding the new foreign policy line includes: periodic efforts by the Dengists to talk about unity not only with the third world but also with all other peace-loving countries and peoples, primarily against aggression and expansion rather than hegemonism in the broadest sense (a formula closer to the "anti-Soviet united front" than the new line); differences in the treatment of foreign policy in the April 1982 draft version of the new state constitution and the final version adopted in December (the "independent foreign policy line" was added); periodic references in nonauthoritative Chinese media to an "international antihegemonist united front" and to the Soviet Union as the "main source of world war," slogans dropped from the authoritative line.

41. *Hongqi* No. 8, 1982, in FBIS May 11, 1982, K4.

5

The Reagan Administration's Southeast Asian Policy

John W. Garver

This chapter analyzes the major contours of the Reagan Administration's policies toward the two regions of Southeast Asia: the communist-ruled states of Vietnam, Laos, and Kampuchea, and the noncommunist states comprising the Association of South East Asian Nations (ASEAN)–Thailand, the Philippines, Malaysia, Singapore, and Indonesia. The relationships between current U.S. policies toward these two regions are analyzed, and these policies compared with those of previous Administrations. An effort is also made to fit these policies into both a broader historic and geographic context.

Twice during the past 40 years the United States has gone to war to prevent Southeast Asia from coming under the control of a power that also dominated the Chinese mainland. In both cases, Indochina was perceived as a stepping-stone to the rest of Southeast Asia, and control of Indochina by the power dominating China was perceived as posing unacceptable threats to vital U.S. interests in the rest of Southeast Asia. The critical Japanese move that set Japan and the U.S. on the collision course that led to Pearl Harbor came in the spring of 1940, when Tokyo decided to take advantage of France's prostration and move into Indochina.[1] It was understood in Washington that Japan's ultimate objective was the oil-rich Dutch East Indies. Similarly, a major reason why the United States began supporting the French colonialist effort in Indochina in 1950 and then itself assumed the major anticommunist role there from 1954 through 1968 was a conviction that the Vietnamese Communists were creatures of Peking and that their victory would result in Chinese control over first Indochina and then over much of Southeast Asia. U.S. decision makers perceived the Vietnamese communists as clients or proxies of the Chinese communists and believed that Chinese domination of Indochina would lead to pro-Chinese communist takeovers in the remaining states of Southeast Asia like a chain of falling dominos.

The Reagan Administration's Southeast Asian policy is characterized by both striking similarities and apparent discontinuities with this "traditional" U.S. policy toward Southeast Asia. The similarities concern the magnitude of U.S. interests in non-Indochinese Southeast Asia, and the linkage between Indochina and the rest of Southeast Asia. At the end of the chapter we will turn to a consideration of the discontinuities between Reagan's and the "traditional" U.S. Southeast Asian policy. Let us begin by considering the similarities.

U.S. INTERESTS IN SOUTHEAST ASIA

Like previous U.S. Administrations, the Reagan Administration perceives vital U.S. interests at stake in the ASEAN area of Southeast Asia. Economic interests are the most tangible. In 1980, trade between the United States and ASEAN countries totaled $21 billion. If ASEAN were considered as a single nation it would constitute the United States' fifth largest trading partner.[2] A breakdown of U.S. trade by regions is presented in Table 5-1.

As Table 5-1 indicates, U.S. trade with ASEAN represents 4% of total U.S. trade. This is considerably less than U.S. trade with North America, Western Europe, Japan, Africa, or Latin America, but still represents a significant block of U.S. trade.

American companies increasingly find the ASEAN countries attractive sites for investment, and U.S. direct investment there in 1981 totaled $4.5 billion.[3] This was a 48% increase over the level of U.S. investment in ASEAN in 1977. Table 5-2 presents figures with regard to the magnitude of U.S. direct investment in ASEAN and figures designed to indicate the size of this investment in relation to that in other regions.

U.S. direct investment in the ASEAN countries represents 2% of all such investment worldwide and the income to U.S. firms from this investment represents 4% of all income on U.S. direct foreign investments. Compared to Latin America or to Central America and Mexico, U.S. direct investment in ASEAN is only 11 or 48% of the totals in those regions. In terms of revenues, however, ASEAN represents 23% and 116% of the income from direct investment in those two other regions. As these figures imply, investment in ASEAN is more profitable than investment in Latin America, Central America and Mexico, or the world generally. The ratio of income on direct investment to direct investment position is 28% for ASEAN, 13% for Latin America, 12% for Central America and Mexico,

TABLE 5-1. U.S. Merchandise Exports and Imports in 1979 by Region of the World (Values in Millions $U.S.)

Region	Exports Value	% of total	Imports Value	% of total	Total U.S. Trade Value	% of total
North America	47,983	26	55,410	27	103,393	27
Western Europe	54,331	30	41,684	20	96,015	25
Asia, excluding Japan, communist countries, and ASEAN	22,691	12	30,578	15	53,269	14
Japan	17,579	10	26,243	13	43,822	11
Africa	6,292	3	24,377	12	30,669	8
South America	13,569	7	13,172	6	26,741	7
ASEAN	6,776	4	9,323	5	16,099	4
Communist countries	7,408	4	2,461	1	9,869	3
Oceania	4,319	2	3,072	1	7,391	2
World	181,802	98[a]	206,327	100	388,129	101[a]

[a]Column does not total 100 because of rounding.

Source: Statistical Abstract of the United States, 1980. U.S. Department of Commerce, Bureau of the Census, 1981, pp. 874–877.

and 13% on a global basis. In other words, ASEAN is a good place to do business.

ASEAN is also an area rich in mineral resources. Four of its five members are on the list of 60 major producers of mineral commodities published by the U.S. Bureau of Mines.[4] Indonesia ranks 15th on this list, accounting for 1.89% of the total value of global mineral production in 1978. Malaysia ranks 30th, just below Brazil and above Chile, producing 0.39% of the world's total mineral wealth. (Incidentally, none of the countries of Indochina are on this list of 60 significant mineral producers.) Indonesia is the world's 14th largest producer of crude petroleum, supplying Japan with 23% and the United States with 6% of their oil imports.[5]

The stability of this oil supply is a major asset to the West. Indonesia refused to participate in the OPEC boycott in 1973, and during the 1980 oil shortage it carried out its supply contracts with U.S. suppliers, while

TABLE 5-2. U.S. Direct Investment in the ASEAN Countries in 1977 and the Relative Significance of this Investment (Values in Million $U.S., Unless Percentages)

	Total assets	U.S. direct investment position	Direct investment income
Indonesia	3,619	984	596
Philippines	5,480	837	91
Malaysia	1,399	464	45
Singapore	10,121	516	97
Thailand	1,205	237	21
ASEAN total	21,824	3,038	850
ASEAN/Latin America	13%	11%	23%
ASEAN/Central America and Mexico	104%	48%	116%
ASEAN/world total	3%	2%	4%

Source: US Direct Investment Abroad, 1977, U.S. Department of Commerce, Bureau of Economic Analysis, April 1981, pp. 6–16. Direct investment is defined as ownership by a single U.S. "person," including a corporation or company, of at least 10% of a foreign business enterprise. Total assets refers to the sum total of owner's equity of affiliates in which there is direct U.S. investment. The direct investment position is defined as the toal equity of U.S. parent firms in, and their net outstanding loans to, their foreign affiliates. Direct investment income is the return to U.S. parent firms on their debt and equity investment in foreign affiliates, i.e., earnings less taxes on dividends and interest.

some other oil producers were selling only on the spot market.[6] Indonesia also produces major quantities of natural gas, tin, and rubber.[7] Malaysia is the world's largest producer of tin, marketing 35% of the world's total tin supply in 1980[8] It also exports substantial quantities of petroleum, copper, bauxite, and tantalum in addition to rubber.[9] Copper, gold, and nickel are among the Philippines' top ten export earners, with substantial quantities of cobalt and chromite also being exported.[10] While not blessed by a super abundance of any single mineral, the Philippines has substantial quantities of several. It is the world's 4th largest producer of chromite, the 9th largest producer of copper, the 6th largest gold producer, and the 5th largest nickel producer.[11] Thailand is a major supplier of tin and also produces tungsten, columbium, and tantalum, as well as rubber.[12]

The ready availability of these mineral resources to the West, and their continued availability during times of crisis or war, is an important concern of U.S. policy. Moreover, the mines and wells of the ASEAN countries are major suppliers of the minerals necessary for Japan's industry.[13] The governance of the ASEAN countries by pro-Western, market-oriented elites is, therefore, an important basis of Japan's economic prosperity. The stability of the Japanese–American alliance is, in turn, tied to Japan's continuing economic prosperity. It follows that the security and stability of the ASEAN countries is linked to the basic structure of U.S. power in East Asia.

Geopolitical considerations also figure in U.S. interests in Southeast Asia. The Indonesian archipelago, stretching for some 3,000 miles, and Malaysia sit astride a series of relatively narrow straits, the control of which could be used to monitor or interdict the passage of warships or merchant vessels moving between the Western Pacific and the Indian Ocean. In the words of the Subcommittee on Asia and Pacific Affairs of the U.S. House of Representatives, "Indonesia is...located in a key strategic position–between the Indian and Pacific Ocean. Our Seventh Fleet must pass through the sea lanes surrounding Indonesia on its way from the Western Pacific to the Persian Gulf, just as Persian Gulf oil passes through these straits en route to South Korea and Japan."[14] Only a few of the myriad channels through the Indonesia archipelago are wide and deep enough to permit the safe passage of submerged submarines. These are: Sunda Strait with a governing depth of 120 feet and a minimum width of 12 nautical miles, Lombok Straits with a depth of 600 feet and width of 11 nautical miles, the Ombai-Wetar Straits with a depth of 600 feet and width of 12 nautical miles, and the Straits of Malacca with a depth of 75 feet and minimum width of 8 nautical miles.[15] These channels are depicted in Figure 5-1.

Several of the ten U.S. Lafayette-class ballistic missile-launching submarines, operating out of Guam, routinely pass through these straits on the way to stations in the Indian Ocean. So do the Trident-class missile submarines, which joined the Seventh Fleet in late 1982.[16] Because the Soviet Union lags behind the United States in antisubmarine warfare (ASW) capability, these straits are important as choke points where the Soviets have some chance of locating transiting U.S. submarines through the use of accoustic devices. They would also be choice spots for Soviet attack, or for hunter–killer submarines to lie in wait for Western oil tankers coming to or from the Persian Gulf or submarines or other warships moving to and from the Indian Ocean. Control of these straits by a friendly government is essential to thwarting Soviet ASW measures while maintaining effective U.S. ASW systems. The importance in this regard of the political sympathies of the Indonesian government was illustrated in February 1982, when Indonesian authorities broke up a Soviet spy ring collecting information that was essential to countering U.S. ASW measures in several of the Indonesian straits.[17]

Beyond these "narrow" economic and strategic interests, the United States has a broader interest in the evolution of the ASEAN countries toward economically developed capitalist democracies. Just as most great powers desire to propagate their values around the world, so too the United States hopes to encourage the ASEAN countries to evolve toward economically prosperous democratic capitalist regimes. This was alluded to by U.S. Assistant Secretary of State for East Asia and the Pacific John H. Holdridge, when he told the Asian Affairs Subcommittee of the

FIGURE 5–1. Strategic Straits and U.S. Military Bases in the Southeast Asian Region.

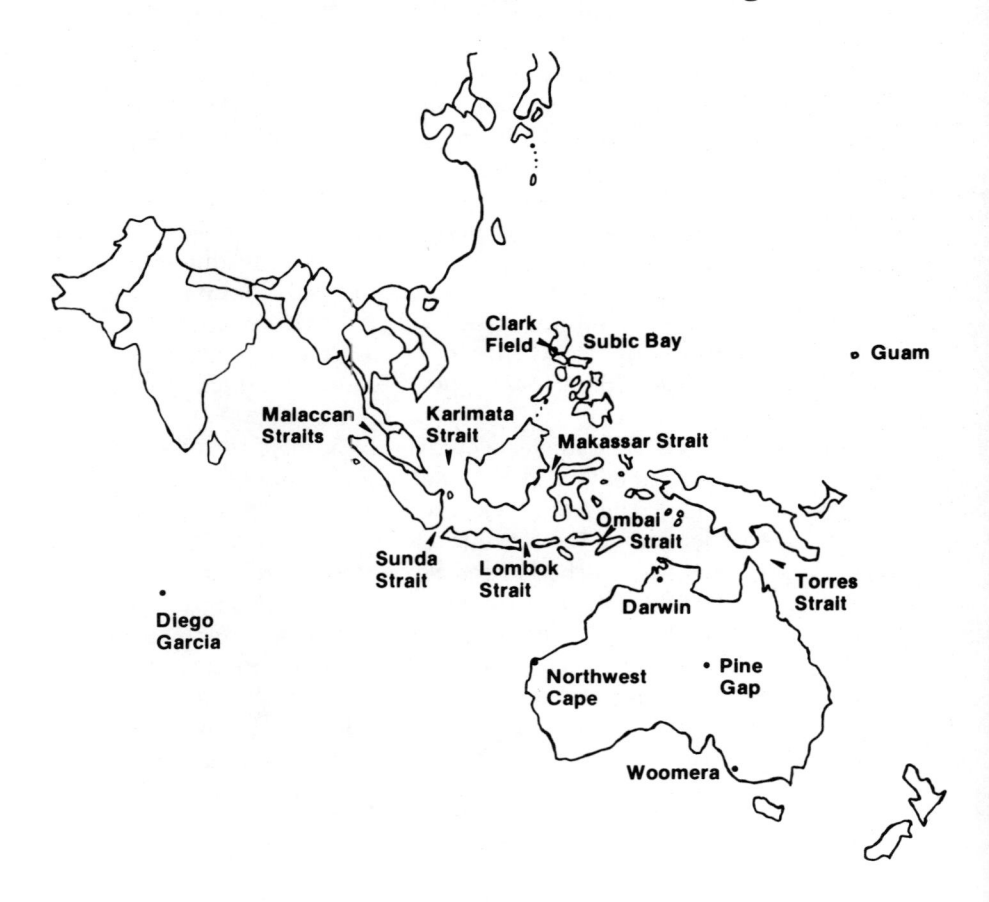

Senate Foreign Relations Committee in July 1981 that maintaining the political stability and economic progress of the "free market" systems of the ASEAN countries lies at the "heart of U.S. policy toward the entire region." According to Holdridge, the five nations of ASEAN, with a population of 250 million, "share a basic pro-Western political and philosophical orientation."[18] As Table 5-3 indicates, U.S. hopes for the development of at least some of the ASEAN countries are not without a basis. The ASEAN countries have achieved a collective annual average rate of growth in real GNP of 5.5% over the 14 years from 1966 to 1980.[19] This is quite a respectable growth rate, compared to other regions of the developing world. Although Indonesia, Thailand, and the Philippines are still poor (in 1979 they had per capita GNPs of US$390, US$590, and US$600, respectively[20]), they have made real progress in development—and have done so under market-oriented economies.

TABLE 5-3. Average Annual Growth in Real GDN, 1966–1980

Indonesia	5.7
Singapore	6.9
Philippines	5.9
Malaysia	4.5
Thailand	4.7
ASEAN Average	5.54

Source: International Monetary Fund, *International Financial Statistics, Yearbook, 1981.* Malaysia average covers period 1970–1980 only.

INDOCHINA AND THE ASEAN COUNTRIES

While U.S. interests in the ASEAN countries are quite substantial, direct U.S. interests in Indochina are few. Since it abandoned its efforts in 1973 and 1975 to prevent communist takeovers of South Vietnam, Laos and Kampuchea, the United States has recognized few interests within Indochina. Perceived U.S. interests which are intrinsic to Indochina, such as full accounting of U.S. soldiers missing-in-action from the pre-1973 period, the possibility of American prisoners being held in Laos or Vietnam, and humanitarian concerns for the peoples of Indochina, are secondary. Rather, U.S. policy toward Indochina is largely a function of the impact of events in that region on the rest of Southeast Asia. Once again,

Indochina is seen as a stepping stone to the rest of Southeast Asia, and U.S. policy in Southeast Asia is largely a function of that linkage. According to Assistant Secretary of State John Holdridge, the Reagan Administration "is convinced that the overwhelming preponderance of U.S. interests in Southeast Asia lie in the ASEAN countries and that our efforts should be toward strengthening bilateral ties with those countries as well as with the organization itself."[21] According to Holdridge, "by any yardstick—population, economic size and dynamism, social and political values, strategic location—the United States has great interests in the five nations of ASEAN." "Our concern with Vietnam," he added, "is a function of the threat which Vietnam poses to ASEAN through its aggression in Kampuchea and through its relationship with the Soviet Union."[22]

U.S. policy makers see several key linkages between Indochina and the security of the ASEAN countries. These include:

The threat posed by a Soviet military presence in Vietnam, especially by major naval facilities at Cam Rahn Bay;

the threat to Thailand posed by Vietnam and by the Vietnamese occupation of Kampuchea;

the danger of Vietnamese and/or Soviet support for communist insurgencies within the ASEAN countries;

the destabilizing impact on the ASEAN countries of large numbers of refugees from Indochina.

U.S. policy in Southeast Asia is designed to meet the challenges posed by these linkages and to insure the continued stability and security of the ASEAN countries.

Soviet warships began operating out of Cam Rahn Bay shortly after the February 1979 Sino–Vietnam War, and by early 1980 Soviet warships and planes were routinely using Cam Rahn Bay and the Danang airfield.[23] This growing Soviet naval presence could have several adverse consequences for U.S. interests in Southeast Asia. It could pursuade various ASEAN governments that they should not risk antagonizing Moscow by doing such things as aiding the anti-Vietnamese guerrilla forces in Kampuchea, hosting U.S. warships or personnel, or supporting U.S. diplomatic initiatives. The next step might well be the establishment of anti-U.S. governments more amenable to Soviet influence. Indonesia and Malaysia view China as posing the greatest long-term threat to their security and might well see some utility in friendship with Peking's great nemesis, the Soviet Union.

The most immediate threat to ASEAN security is posed, Washington believes, by the Vietnamese occupation of Kampuchea. A Vietnamese victory in a war with Thailand would have profound consequences on U.S.

power in the region. The utility of friendship with the United States would be drawn into question and a neutralist government would probably come to power in Bangkok. In a congressional testimony in March 1981, Michael H. Armacost, Director of the State Department's Bureau of East Asian and Pacific Affairs, enumerated Vietnamese incursions against Khmer guerrilla encampments inside Thailand and reconnaissance patrols into Thailand as the major Vietnamese threats to Thai security. Because of the "increased military threat from Vietnam" Armacost said, "the Thai military needed force improvements to present a plausible deterrent to Vietnamese forces."[24]

Less stressed by Washington, but probably in fact recognized as of greater potential danger than conventional military aggression by Vietnam, is the threat that would be posed by Vietnamese support for indigenous insurgent forces in various ASEAN countries. The "domino theory" in various forms shaped U.S. policy in Southeast Asia for several decades. The fear of guerrilla insurgencies on which that theory was based is especially strong in the Reagan Administration. All of the ASEAN countries have faced strong and tenacious Communist-led insurgencies or movements during the post-WW II era; and Thailand, the Philippines, and Malaysia currently confront such insurgencies. Thailand and the Philippines especially are the most likely candidates for future "dominos" in Southeast Asia. Thailand's northern provinces, northeast, and southern panhandle all have significant guerrilla forces led by the Communist Party of Thailand (CPT) and challenging the government for control of these regions.[25] If Vietnam were to begin supporting these insurgent forces across the long Thai–Laos and Thai–Kampuchean borders, Bangkok's security problems would be greatly magnified. Although since 1975 Hanoi has shown no inclination to subvert the Thai government, both Bangkok and Washington are concerned that if Hanoi did decide to to this, its domination of Kampuchea would put Hanoi in a strong position. Conversely, the creation of a neutral Kampuchea not under Vietnam's domination and not occupied by Vietnamese troops would be a major plus for Thai security.

The Philippines is perhaps the prime choice for domino status. It is characterized by extreme and growing economic inequality, massive poverty, the alienation of much of the middle classes from a corrupt and dictatorial government, increasing urban terrorism, and strong communist (3,000–5,000 armed men) and secessionist (5,000–9,000 armed men) movements.[26] Again, Vietnam is not supporting these antigovernment insurgencies, but Washington probably fears that it may decide to. If Vietnamese aggression in Kampuchea is successful, it is feared that Hanoi's next move may be to subvert its most vulnerable neighbors, especially one such as the Philippines, which hosts a major U.S. military presence. One of the premises underlying U.S. policy in Southeast Asia is that aggression unchallenged becomes aggression emboldened.

While considering the domestic security problems of the ASEAN countries, it is important to understand that several of these countries have traditionally viewed China as the most prominent revolutionary and expansionist power in Southeast Asia. The Chinese Communist Party (CCP) was long the main external supporter of the CPT, the Malaysian Communist Party, the New People's Army in the Philippines, and the Communist Party of Indonesia, and frequently supplied money, training, arms, and propaganda support to these revolutionary movements. Although such support drastically declined throughout the 1970s, it continues at a very low level and China has yet to renounce such "party-to-party" attempts to subvert its neighbors.[27] As of mid-1982 the CCP still refused to break relations with these revolutionary movements on the grounds that they were party-to-party ties that were of no legitimate concern to the ASEAN governments.

In the case of Indonesia and Malaysia (and to a lesser extent, Thailand), apprehension of China is further exacerbated by the existence of large, unassimilated, and economically powerful populations of overseas Chinese—which were also the backbones of the once strong Communist movements in those countries.

U.S. POLICY IN SOUTHEAST ASIA

According to Assistant Secretary of State Holdridge, the United States recognizes three basic "objectives" in Southeast Asia. These include:

1. Maintaining the political stability and economic progress of the "free market" systems of the ASEAN countries. This is the "heart" of U.S. policy toward the entire region.
2. In cooperation with ASEAN, restraining Vietnamese aggression.
3. Curbing the growing Soviet military presence and influence in the region.[28]

These three objectives can be seen as an operationalization of the containment of Soviet influence. Vietnam and the Soviet Union are seen as aggressive and subversive forces posing threats to major U.S. interests in the ASEAN countries. Generally speaking, the Reagan Administration's Southeast Asia policy is based on a perception that since the mid-1970s the United States has become dangerously weak and passive, thereby encouraging the Soviet Union to attempt to expand its influence and control in various areas of the world. To counter this trend the Administration wants to strengthen U.S. alliances, build a strategic consensus with regard to the danger of Soviet expansion, play a more activist role in thwarting the advances of the Soviet Union and its allies, and build up U.S. military power and that of its allies to deter or defeat Soviet and Vietnamese military moves.

This view of a dangerous U.S. weakness and passivity began to emerge during the last 2 years of the Carter Administration. Since the earliest U.S. decisions to withdraw from Vietnam, Washington attempted to assure Asian governments that, however the Vietnam War might end, the United States did not intend to withdraw from Asia and was determined to remain a major actor in East Asia. But events and U.S. behavior often seemed to belie these assurances. As Table 5-4 indicates, there was a gradual decline in U.S. troop deployments in East Asia throughout the latter half of the 1970s.

Coming shortly after the dramatic collapse of the pro-U.S. regimes in Indochina, the 1975–1976 withdrawal of the 23,000 U.S. troops in Thailand at the request of the Thai government seemed a harbinger of major shifts in the Southeast Asian balance of power.[29] The discussion in the United States about the same time of abandoning the U.S. bases in the Philippines in favor of bases in the Marianas Islands, contributed to the perception of U.S. retreat.[30] The Carter Administration's March 1977 decision to withdraw all U.S. ground combat personnel from South Korea was a bombshell that seemed to confirm the most dismal prognoses of U.S. intentions.[31] The formal dissolution of the South East Asian Treaty Organization (SEATO) in February 1976 give further substance to the view that the era of U.S. preeminence in Southeast Asia was past.[32] Throughout this period Soviet naval power in the Pacific was expanding rapidly and Moscow was inching toward a military presence in Vietnam.

The escalating confrontation between China and Vietnam during 1978 led the Carter Administration to reconsider the direction of U.S. policy and to conclude that further reductions in the U.S. political role and military presence in Southeast Asia might have grave consequences. Washington increasingly feared that unless U.S. power in Southeast Asia was maintained, the Sino–Vietnamese conflict might expand to the whole of Southeast Asia, with ASEAN being polarized into pro-Chinese and pro-Vietnamese/Soviet blocs and with Vietnam being tempted to attack or subvert Thailand. As a result of these concerns, the troop withdrawal from South Korea was suspended, discussion of the proposed withdrawal of one carrier task force group from the Seventh Fleet was suspended, greater priority was given to encouraging Japan to rearm, more sympathetic treatment was given to ASEAN requests for weapons purchases, and priority was given to helping Thailand modernize its armed forces.[33]

TABLE 5-4. U.S. Military Personnel in East Asia and the Western Pacific, 1975–1980

	31 March 1975	31 March 1976[a]	31 Dec 1977	30 Sept. 1978	30 Sept. 1979	30 Sept. 1980
Japan (including Okinawa)	51,000	48,000	48,766	45,939	46,207	46,004
South Korea	42,000	41,000	39,964	41,565	39,018	38,780
China:						
People's Republic	–	–	–	–	9	12
Taiwan	4,000	2,000	949	–	1	1
ASEAN						
Philippines	16,000	16,000	14,380	14,433	14,101	13,387
Thailand	23,000	4,000	141	104	102	95
Indonesia	–	–	73	61	71	55
Malaysia	–	–	14	17	13	15
Singapore	–	–	25	26	23	23
Hongkong	–	–	337	38	34	39
Burma	–	–	13	12	11	10
Afloat; Seventh Fleet	22,000	22,000	25,607	25,890	21,910	15,515
Other[b]			3			2
Total East Asia	158,715	134,000	132,059	129,584	122,227	114,845
Guam	10,000	9,000	8,454	8,380	8,756	9,053
Total East Asia and Guam	168,715	143,000	139,165	137,211	130,983	123,896

Source: Office of the Secretary of Defense, *Selected Manpower Statistics,* annual reports for years 1975 through 1980 and fiscal year 1980. American Statistical Index Number 3544-1.

[a]Countries with less than 100 personnel are apparently not listed. Dash indicates no figure given in official list for that year.
[b]Includes three in Laos in 1977 and two in Nauru in 1980.

In order to meet the challenges to U.S. interests posed by Vietnamese and Soviet moves in Southeast Asia, the Reagan Administration has continued a number of policies that emerged during the Carter or Ford Administration. These policies include:

1. Encouraging the development of cooperation among the ASEAN countries and strengthening ASEAN as an organization.
2. Making clearer the U.S. commitment to the security of the ASEAN countries in the expectation that this will deter Soviet or Vietnamese aggression.
3. Increasing pro-Western military capabilities in the region by:
    a. Encouraging Australia and New Zealand to continue their contribution to ASEAN security.
    b. Encouraging the ASEAN countries to increase their own military preparedness by increasing arms sales and military assistance to them.
    c. Strengthening the U.S. military presence in the Southeast Asian region.
4. Reducing the disruptive impact upon the ASEAN countries of refugees from Indochina.
5. Cooperating with China to jointly guarantee the security of Thailand.
6. Cooperating with China and ASEAN to force Vietnam to withdraw its troops from Kampuchea.

Support for ASEAN as an organization is a key element of Reagan's Southeast Asian policy. The United States, of course, has supported and encouraged ASEAN since its formation in 1967, but since the collapse of the pro-U.S. regimes in Indochina in 1975 ASEAN has assumed increased importance. In the words of Assistant Secretary Holdridge:

> The aggressive behavior of Vietnam since 1975 has given great impetus to [ASEAN] solidarity, and the organization has become a significant force in world politics. U.S. policy has been and will continue to encourage this trend.[34]

Increased cooperation among the ASEAN countries has several advantages from the U.S. perspective. Most broadly, a cohesive ASEAN supported by and friendly to the United States, Australia, New Zealand, and other Western countries is an effective way of excluding a significant Soviet political or military presence from the region. (A rough historical analogy might be British support for the Ottoman Empire in the 19th century as a way of checking Russian advances.) Again according to Holdridge, "positive, active support for ASEAN is the most effective means of curbing the ambitions of Vietnam and the Soviet Union."[35] Firm support by its ASEAN partners will encourage each country to accept higher risks in opposing Vietnamese aggression. For example, Thailand decided to support the Khmer Rouge resistance after Vietnam's 1978 invasion of Kampuchea, in part because of the strong and swift support of its ASEAN partners.[36] Diplomatic initiatives endorsed by a united ASEAN are likely to win broader support than those sponsored mainly by a single member or by the United States.

It has been ASEAN, for example, that has persuaded the United Nations General Assembly (UNGA) to continue seating the Democratic Kampuchean regime – a feat that neither the United States nor China would have been likely to accomplish. As Holdridge told the Senate Foreign Relations Committee in July 1981, "a unified stance on Kampuchea at the United Nations and elsewhere has greatly strengthened ASEAN's hand against Vietnam and has preserved for it the diplomatic initiative."[37]

A cohesive ASEAN is also one component of a broader U.S. Pacific policy. Together with Australia and New Zealand in the South Pacific and with Japan in the North Pacific, a bloc of relatively prosperous and democratic capitalist states in Southeast Asia would help form a peaceful and interdependent "Pacific Community" during the latter part of the 20th century.[38]

Increased cooperation among the ASEAN countries also has security payoffs. At the first ASEAN summit conference in Bali, Indonesia, in February–March 1976, the idea of ASEAN becoming a mutual security organization was rejected, but bilateral security ties among its members were to be developed "in accordance with their mutual needs and interests."[39] The Bali conference was a major step forward in ASEAN unity. It established an ASEAN secretariat and set up a ministerial-level council to resolve political disputes. The goal of the secretariat was defined as turning ASEAN into a zone of peace, freedom, and neutrality. Since 1976 ASEAN has continued to eschew a formal collective security arrangement, while steadily and gradually expanding bilateral security cooperation.[40] The elimination of residual territorial conflicts among the ASEAN countries also reduces sources of conflict that might be exploited by Vietnam or the Soviet Union or hinder further bilateral security cooperation.[41]

As the Sino–Vietnamese conflict escalated during 1978, Washington attempted to clarify its concern for and commitment to the Southeast Asian region, especially Thailand, in hopes that this would deter Vietnamese or Soviet adventurism. In early 1978 Vice-President Mondale visited the Philippines, Thailand, Indonesia, Australia, and New Zealand in an attempt to underline U.S. interests in Southeast Asia.[42] Mondale's visit to Thailand symbolized the improvement in Thai-U.S. relations that had begun with the rise of Kriangsak Chamanand to power in a military coup in October 1977.[43] While in Bangkok, Mondale reaffirmed the U.S. pledge of 1954 to help defend Thailand's security.[44]

Vietnam's invasion of Kampuchea caused radical changes in Thailand's strategic perspective. Since 1975 Thailand's various governments had pursued a policy of détente with the Asian Communist powers, including both China and Vietnam, although China was perceived as standing behind the two major threats to Thai security—the insurgency led by the China-supported CPT, and Thailand's Chinese minority.[45] After 1975 the CPT had reoriented its activity from the border area adjacent to Laos to areas along Thailand's borders with Kampuchea. Since Laos was Vietnam's client while Kampuchea then enjoyed Chinese support, Bangkok saw this reorientation as an indication of declining Vietnamese and increasing Chinese support for the CPT. Moreover, China continued to support the CPT; it was not until mid-1979, for example, that the CPT-operated radio station in South China was shut down. Once Kampuchea came under Vietnamese occupation, however, Bangkok perceived Hanoi as now posing a greater danger to Thailand than did Peking.

This shift in strategic outlook brought Prime Minister Kriangsak to Washington in February 1979 to seek firmer U.S. assurances of its concern for Thai security. This was the first visit by a Thai prime minister since Kittikachorn visited in May 1968,[46] and led to the revitalization of the Thai–U.S. security relationship that had been virtually moribund since the traumatic expulsion of U.S. troops from Thailand in 1975–1976. Washington was very responsive to Bangkok's new security concerns in the wake of Vietnam's takeover of Kampuchea, and provided strong guarantees to Thailand. In greeting Kriangsak, President Carter said that the United States was "intensely interested" in and "deeply committed" to the integrity of Thailand's borders. During their discussions, Carter promised Kriangsak that the United States would take "definite action" on behalf of Thai security should the latter be threatened.[47] Such statements of U.S. support continued throughout 1979 and 1980, and stepped-up shipments of U.S. arms to Thailand (discussed below) and visits by elements of the Seventh Fleet to Thai ports gave substance to the proclamations.[48]

The Reagan Administration has continued this effort to convince relevant actors of the U.S. commitment to Southeast Asia. Secretary Haig's attendance of the conference of ASEAN foreign ministers in Manila in June 1981 (the third consecutive participation by a U.S. Secretary of State in these annual conferences) was intended to confirm "the great level of support that the United States has for this regional grouping and the importance that the United States attaches to the bilateral relationships it maintains with each of the member nations of ASEAN."[49] Haig also reiterated that the United States would honor its obligations to Thailand under the Manila Treaty and would carry out the letter and spirit of its commitments to the Philippines under their mutual defense treaty. The visit of Singaporean Prime Minister Lee Kuan Yew to Washington in June 1981 was intended, in part, as a demonstration of a strengthening of U.S.-ASEAN ties, as was Vice-President Bush's visit to Singapore in April 1982, as part of his two-week tour of six Asian countries.[50] The visits of Presidents Marcos of the Philippines in September, 1982, and Suharto of Indonesia, in October, 1982, are also indications of Reagan's attempt to strengthen ties with those two countries. This will be the first visit by a Philippine President to the United States since 1969 and the first by an Indonesian leader since 1975. Table 5-5 illustrates top-level ASEAN-U.S. interactions. Figure 2-2 depicts graphically ASEAN top-level visits to the United States and shows that after falling sharply with the end of the Vietnam War, such visits again increased sharply with the intensification of the Sino-Vietnam conflict in 1978.

FIGURE 5-2. Visits by Top-Level Southeast Asian Leaders to the United States since WWII.

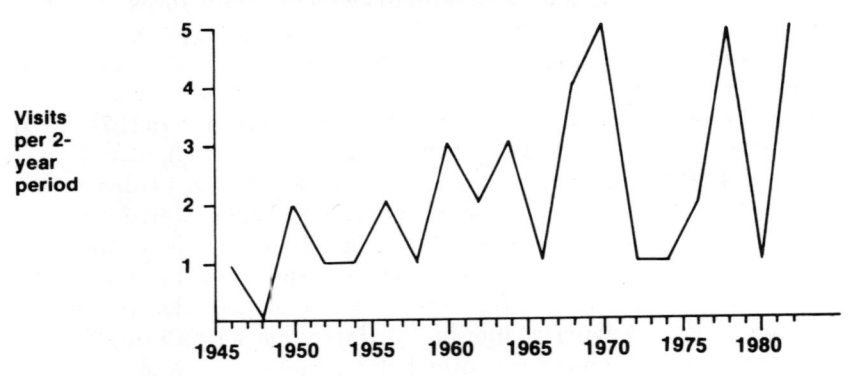

A significant nuance in Reagan's support for Washington's ASEAN friends has been the deemphasis on the human rights violations of pro-U.S. authoritarian regimes, especially the government of Ferdinand Marcos in the Philippines. Human rights were a major issue in Philippine-U.S. relations under Carter. In January 1977, for example, Marcos threatened to terminate the military base agreement with the United States because of U.S. criticisms of human rights violations in the Philippines.[51] During Mondale's visit to the Philippines in May 1978, he met with various opposition leaders and publicly spoke of U.S. interest in promoting democracy and individual liberty in the Philippines, much to the annoyance of Marcos.[52] The Reagan Administration, however, has decided not to let human rights problems estrange U.S. relations with the Philippines. This was symbolized by Vice-President Bush's attendance of President Marcos' inauguration in Manila on June 30, 1981. The presidential

TABLE 5-5. Top-Level Visits between the ASEAN Countries and the United States during the Post WWII Period

Philippines
President-elect Roxas	May	1946	President Eisenhower	Jun.	1960[a]
President Quirino	Aug.	1949	President Johnson	Oct.	1966
	Feb.	1950	President Nixon	Jul.	1969
	Aug.	1951	President Ford	Dec.	1975
	Aug.	1953			
President Garcia	Jun.	1958			
President Macapagal	Nov.	1963			
	Oct.	1964			
President Marcos	Sept.	1966			
	Mar.	1969			
	late	1982[b]			

Thailand
PM Pibulsonggram	May	1955	President Johnson	Oct.	1966
King Bhumibol	Jun.	1960		Dec.	1967
	Jun.	1967			
PM Kittikachorn	Mar.	1968	President Nixon	Jul.	1969
PM Kriangsak	Feb.	1979			
PM Prem	Oct.	1981			

Malaysian
PM Rahman (Malaya)	Oct.	1960	President Johnson	Oct.	1966
	Jul.	1964			
	Oct.	1969			
PM Razak	Oct.	1971			
PM Hussein	Sep.	1977			
King and Queen	Mar.	1978			

Indonesian
President Sukarno	May	1956	President Nixon	Jul.	1969
	Oct.	1960	President Ford	Dec.	1975
	Apr.	1961			
	Sep.	1961			
President Suharto	May	1970			
	Jul.	1975			
	Fall	1982[b]			

Singapore
PM Lee	Oct.	1967	May	1975	None	
	Dec.	1968	Oct.	1977		
	May	1969	Jan.	1978		
	Nov.	1970	Oct.	1978		
	Apr.	1973	Jun.	1981		
	May	1975	Jul.	1982		

Sources: Office of the Historian, Bureau of Public Affairs, U.S. Department of State, *Lists of Visits of Foreign Chiefs of State and Heads of Government to the United States 1789–1978*, January 1979, Research Project No. 495A, 12th Revision; "Presidents Abroad," *Department of State Bulletin*, Sept. 1981, Vol. 81, No. 2054; Memorandum from the Office of Protocol to author, 12 July 1982, "Visits by ASEAN Officials (Foreign Minister and above-ranking Officials) Since 1978 – Per Protocol Records." US Department of State.

[a]This was the first visit by a U.S. President to an East Asian country.
[b]Confirmed by calls to the country embassies in Washington, D.C.

election of 1981 was the first to be held in the Philippines since the proclamation of martial law in 1972, but was boycotted by the opposition because they felt it was so restricted as to be fraudulent. Nevertheless, in his speech Bush lauded Marcos' "adherence to democratic principles and to the democratic process" and clearly associated the United States with President Marcos when he said that "We stand with the Philippines; we stand with you, sir."[53]

Another major component of U.S. policy in Southeast Asia is to strengthen the regional military capabilities of the United States and its friends and allies. The major U.S. military bases in the Southeast Asian region are on the U.S. territory of Guam Island and Clark Air Force base and Subic Naval base in the Philippines. (These locations are shown in Figure 5-1.) The retention of the latter two bases involved major U.S. policy decisions during the Ford and Carter Administrations. For most of the post-WW II era the United States used these bases rent-free under a treaty signed in 1947.[54] Negotiations for a new treaty that provided for U.S. payment for use of the bases began during the Ford Administration. After considering and rejecting proposals to abandon the Philippine bases in favor of bases in the Mariana Islands some 800 miles to the east, the Carter Administration negotiated an agreement, by the end of 1978, which provided for "unhampered" U.S. use of the bases, but which recognized exclusive Philippine authority over base security and ultimate control at each base. In return the United States was to provide the Philippines with $100 million per year for 5 years, one-half in the form of economic assistance and one-half in foreign military sales (FMS) credits.[55] (During the long negotiations over the bases, the Philippines was under pressure from both its ASEAN partners and China to allow the U.S. military presence in the Philippines to continue.) The new bases agreement was signed in January 1979 and was to run through 1999, with provisions for review of compensation in 1984 and every 5 years thereafter.[56] President Marcos became unhappy with certain provisions with regard to base security, however, and requested an early renegotiation of those parts of the treaty. During his visit to the Philippines in April 1982, U.S. Secretary of Defense Caspar Weinberger agreed to the beginning of negotiations on these issues with an eye to reaching an agreement by 1984. President Marcos also discussed the base issue with President Reagan during his official visit to Washington in September 1982.[57]

Both the Carter and Reagan Administrations have expanded the U.S. military presence in Australia in an effort to ensure U.S. control over the seas of Southeast Asia and the Indian Ocean. In May 1980, the United States requested that Australia permit U.S. B-52 bombers operating out of Guam to use Darwin as a staging area for surveillance flights over the Indian Ocean area.[58] These negotiations were concluded under the Reagan

Administration in March 1981, when an agreement was signed allowing B-52s to use the Darwin airfield, but not to carry nuclear weapons into Australia without the prior consent of the Australian government. One hundred U.S. Air Force personnel were also to be stationed at Darwin to service the bombers.[59] During 1980 and 1981, U.S. electronics and communications bases at Pine Gap in central Australia, at Northwest Cape on the coast of western Australia, and at Woomera in southern Australia were also expanded. The base at Pine Gap is involved in satellite surveillance of Soviet targets on land and sea,[60] while the base at Northwest Cape maintains contact with submerged U.S. submarines operating in the Indian and Pacific Oceans.[61] The base at Woomera may be involved in monitoring the movement of ships to the south of Australia. These bases are essential elements in the U.S. effort to counter the growth of Soviet naval power in Southeast Asia and to maintain U.S. naval superiority in those seas.

It is also possible that the United States has contingency plans to use a large (13,000 acres) air base that the Malaysian government is developing at Gong Kedak in the northeastern state of Kelatan.[62] U.S. use of this base, which is to be operational by 1983, would significantly enhance U.S. capabilities in the Gulf of Thailand and the South China Sea.

A major aspect of the U.S. effort to increase military capabilities in Asia is the so-called "multinational strategy" under which friendly governments are encouraged to increase their own military preparedness and cooperation.[63] Japan is perhaps the main object of U.S. efforts in this regard. The United States would like Japan to assume primary responsibilities for sealing up the straits leading out of the Sea of Japan and for patrolling the sea lanes in the arc of sea to the south of Japan, north of the Philippines, and west of Guam.[64] Japan has thus far rejected these proposals, but Washington spends considerable, but low-key, diplomatic pressure prodding Japan in this direction.[65] Were Japan to accept such a role, this would free U.S. naval and air forces for redeployment to the Southeast Asian region and the Indian Ocean.

The United States also encourages Australia and New Zealand to bolster their commitment to the defense of the ASEAN countries. The United States supported the reactivation in the fall of 1980 of the five-power defense arrangement between Malaysia, Singapore, New Zealand, Australia, and Britain.[66] During his participation in the ANZUS conference in June 1981, Secretary of State Haig urged New Zealand to keep its infantry battalion in Singapore, and Australia to keep its two squadrons of Mirage jet fighters stationed in Malaysia.[67]

Another important component of the U.S. multinational strategy is to use increased levels of military assistance to encourage the ASEAN countries to expand their own military capabilities. As the Sino–Vietnam conflict escalated and Vietnam moved into Kampuchea, the ASEAN countries began increasing their defense budgets at a rapid rate. In 1980 total ASEAN defense spending was $5.5 billion, 45% greater than the 1979 level and double that of 1975.[68] The United States is encouraging this trend through an expansion of its military assistance programs. The U.S. view was encapsulated by Secretary of Defense Harold Brown in his annual report of 1981:

> U.S. security assistance will be a key element in advancing U.S. interests and in promoting regional security, especially for South Korea, the Philippines, Thailand, Indonesia, Singapore, Australia, New Zealand, and Malaysia. The benefits of these programs, in terms of greater military capabilities, are shared by all those who have vital interests in the region, including the United States.[69]

Levels of U.S. military assistance to the ASEAN countries are depicted in Table 5-6 and trend lines are illustrated in Figure 5-3.

TABLE 5-6. U.S. Military Assistance to the ASEAN Countries, 1975–1983 (Values in Millions $U.S.)

	1975	1976	1977	1978	1979	1980	1981	1982	1983
Philippines	77.3[a]	36.3[a]	38.1[b]	37.3[b]	31.7[b]	75.5[c]	75.6[c]	52.8[c]	102.1[d]
Singapore	–	–	–	–	–	–	–	0.05	0.05
Thailand	88.1	42.5	47.2	38.6	32.1	40.4	52.9	(73.5)[d]	103.2
Malaysia	5.0	17.3	36.3	17.1	8.0	7.3	10.3	13.2	13.4
Indonesia	21.0	46.0	40.8	58.1	34.8	33.1	32.0	48.1	52.1
ASEAN Total	191.4	142.1	162.4	151.1	106.6	156.3	170.8	187.6	270.8

Source: Each column is from following sources,

[a]*Statistical Abstract of the U.S., 1978*, U.S. Department of Commerce, Bureau of the Census, pp. 872–873. Years end June 30 except for 1976, which ends September 30.

[b]*Statistical Abstract of the U.S., 1980*, pp. 870–871. Years end June 30.

[c]*Foreign Assistance Legislation for FY 1982 (Part 5)*, Hearings before the Subcommittee on Asian and Pacific Affairs, U.S. House of Representatives, 97th Congress, 1st Session, March–April 1981, pp. xv–xxiv. Figures for 1980 are actual, 1981 estimated, and for 1982 are the Reagan Administration's request and the Subcommittee's recommendation.

[d]Foreign Assistance Legislation for FY 1983, draft of subcommittee print obtained from subcommittee, July 1982. Figures are Administration's request and recommendations of Asian and Pacific Affairs Subcommittee of the Foreign Affairs Committee of the House of Representatives.

U.S. military assistance to Thailand has shown the most rapid and consistent growth. In the U.S. view, Thailand's "vulnerability as a frontline state in the continuing conflict in Southeast Asia, as well as its acute

FIGURE 5-3. U.S. Military Assistance to the ASEAN Countries.

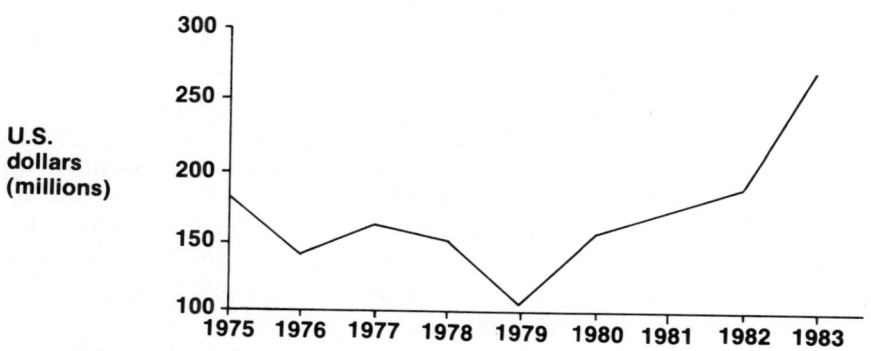

development needs" has made imperative increases in both military assistance and economic aid.[70] U.S. assistance to Thailand is seen as a way of developing a credible deterrent to further Vietnamese aggression against Thailand, to provide a symbol of U.S. support for Thailand, and of increasing the government's ability to repress domestic insurrection. The deterrent and symbolic aspects were clearly illustrated during 1979 and 1980. When Prime Minister Kriangsak visited Washington shortly after Vietnam's invasion of Kampuchea and the United States wished to demonstrate firm support for Thailand, President Carter assured Kriangsak of a continued supply of weapons and agreed to speed up the delivery of military equipment already purchased by Thailand.[71] Then when heavy fighting between the Khmer Rouge and Vietnamese forces erupted near the Thai–Kampuchean border in January 1980, the United States once again agreed to accelerate the delivery of military supplies to Thailand.[72] And after 2,000 Vietnamese soldiers made raids into Thailand for about 10 hours on June 23, 1980, the United States agreed to speed up the sea delivery of 35 M48A5 tanks to Thailand, to undertake an emergency airlift to Thailand of some $3 million worth of small arms and other previously ordered equipment, and to begin procedures to allow Thailand to borrow funds to purchase additional military equipment.[73]

The Reagan Administration has further increased military assistance to Thailand. The increases requested under the Administration's first budget, from $53 million to $82 million, represented the largest increase for any East Asian country. This amount included $50 million in the form of direct credits at a concessional rate, Thailand being the first East Asian country to receive such direct credits.[74] These funds were later pruned by Congress, but the $74 million spent still represented a hefty 39% increase. For fiscal year 1983 the Administration has requested a further 40% increase to $103 million.[74]

Both Carter's and Reagan's policies of stepping up support for Thailand are in line with the use of Thailand as a base from which to pressure Vietnam into withdrawing its troops from Kampuchea; adequate deterrence is necessary to prevent Vietnam from moves into Thailand to destroy the base areas used by the Khmer resistance forces.

The other substantial increase in military assistance to the ASEAN countries under Reagan's first budget was to Indonesia. Here a 50% increase from $30 million to $45 million was justified by the Administration by increasing Soviet and Vietnamese threats and Indonesia's strategic position astride sea lanes between the Indian and Pacific Oceans.[75] An unmentioned consideration was probably the desire to provide Indonesia with additional incentives not to break ranks with its ASEAN partners over the issue of pressuring Vietnam.

A major aspect of U.S. Southeast Asian policy has dealt with refugees from Indochina. Up to the fall of 1981, the United States had taken in nearly 650,000 refugees from Indochina, while also urging other countries to accept substantial numbers.[76] The United States has also spent substantial funds to feed and house Indochinese refugees in various Southeast Asian countries. Again the Reagan Administration has continued this policy. For fiscal year 1982 the Administration requested $30 million for assistance for various international programs to meet the needs of Indochinese refugees located in ASEAN countries and Hongkong,[77] while urging the maintenance of relatively high quotas for immigration of Indochinese refugees (120,000 during FY 1982, as compared to 168,000 in FY 1981).[78]

There are several considerations underlying U.S. policy with regard to Indochinese refugees. These are: (1) humanitarian concerns; (2) maintaining the stability of the ASEAN countries; (3) keeping the pressure on Vietnam to withdraw from Kampuchea.

The first two points will be briefly discussed here; the third point is discussed in the section below on pressuring Vietnam. Compassion for the tragic plight of the Khmer people and for Vietnamese for whom the United States fought so long is without question a major concern of U.S. refugee policy. This was especially true during the Carter Administration, which emphasized its commitment to human rights and morality, and which held office during the massive outpouring of refugees from Kampuchea and Vietnam in 1978–1980. One of the reasons why the United States accepted such large numbers of refugees from Indochina is that the alternative solution of letting the ASEAN countries turn away refugees either by towing their boats back out to sea, or in the case of Thailand, by herding them across the Thai–Khmer border, was unacceptable to Washington largely for humanitarian reasons. Such a spectacle would also have eroded support by the American public for the ASEAN countries involved in such practices, an instance of moral considerations acting through political ones.

U.S. refugee policy is also concerned with the stability of the ASEAN countries. In testimony before the Senate Judiciary Committee in September 1981, Under Secretary of State Walter J. Stoessel explained that large populations of refugees would "threaten the stability of countries where they seek first asylum."[79] Such unwanted and uprooted people would provide recruits for revolutionary or terrorist movements; they might become the Palestinians of Southeast Asia, thereby adding considerably to the internal security problems of Malaysia or Thailand. Moreover, because large numbers of the refugees from Vietnam are overseas Chinese, their arrival in Indonesia and Malaysia could exacerbate the anti-Chinese sentiments common in those countries. Increased communal strife might in turn increase political instability, retard economic development, or radicalize segments of the population. It could also generate popular pressure on the government to adopt more "anti-Chinese" foreign policies, including policies more sympathetic toward Vietnam.

U.S. STRATEGIC ENTENTE WITH CHINA AND SOUTHEAST ASIAN POLICY

During both the Carter and Reagan Administrations, U.S. efforts to contain Vietnamese and Soviet influence in Indochina and Southeast Asia have been linked to the strategic partnership with China. Since the end of 1978, China and the United States have actively cooperated to isolate Vietnam, to pressure it into withdrawing from Kampuchea, and to prevent it from attacking other countries in Southeast Asia, especially Thailand.

Within the Carter Administration, there were divergent views about policy toward China and Vietnam. Secretary of State Cyrus Vance advocated a go-slow approach toward developing security relations with China, while National Security Advisor Zbigniew Brzezinski favored a rapid normalization of relations and expanded cooperation to help contain Soviet initiatives. By early 1978, Brzezinski had concluded that Vietnam was acting as a Soviet "proxy" in its conflict with Kampuchea, and he probably urged joint Sino–U.S. efforts to contain Vietnam. In June 1978, Brzezinski was dispatched to China to explore the possibilities of expanded anti-Soviet cooperation, among other matters, and on his return, President Carter decided to move ahead with the normalization of Sino–U.S. relations in 1978.[80] Vance was opposed to the movement toward a strategic relationship with China, and along with Assistant Secretary of State for

East Asia and the Pacific Richard Holbrooke, he urged a more or less simultaneous normalization of relations with both Vietnam and China. Vance realized that the normalization of relations and expanding security cooperation with China, in the context of continued U.S.–Vietnamese estrangement, would inevitably create a Sino–U.S. bloc against Vietnam, behind whom would stand the Soviet Union. In other words, it would contribute to the polarization of world politics and the deterioration of Soviet–U.S. relations, which Vance hoped to avoid.

Negotiations between Hanoi and Washington and between Peking and Washington continued simultaneously throughout the summer of 1978. In July, Hanoi dropped its previous demand for reparations from the United States, and by September, Vietnam and the United States had reached virtual agreement on the terms of normalization. The conclusion of the 30-year treaty of friendship and mutual security between Vietnam and the Soviet Union in November, however, swung the balance within the U.S. government toward Brzezinski's point of view, and killed any possibility of U.S.–Vietnam normalization for the time being.[81] By the end of 1978, U.S. intelligence reports made it clear that Vietnam was preparing to invade Kampuchea and the conclusion of the Soviet–Vietnam treaty was taken as proof that Vietnam acted with Soviet blessings and as a Soviet proxy in Indochina. Events in Indochina had decisively shifted the factional balance in Washington, and the United States moved to join with China to contain Vietnamese expansion in Indochina.

The United States tacitly supported China's punitive war against Vietnam in February–March 1979. U.S. leaders did not warn against clear threats of such a move by Deng Xiaoping while he was in the United States in January 1979, and during the fighting Washington called for both a Chinese withdrawal from Vietnam and a Vietnamese withdrawal from Kampuchea, which was exactly what Peking was attempting to achieve. China was far from happy with the lukewarm nature of U.S. support during the war, but Washington clearly intended its actions to constitute a "tilt" toward China during that brief conflict.[82]

Throughout 1979 and 1980, both China and the United States sought to mobilize political and economic pressures on Vietnam to compel it to withdraw from Kampuchea. Both supported ASEAN's efforts to pass resolutions in the U.N. General Assembly calling for Vietnam's withdrawal from Kampuchea and to continue seating Democratic Kampuchea as Kampuchea's representative in the UNGA. This combined ASEAN–U.S.–China bloc has been successful in winning broad support in the UNGA, leaving Vietnam with support only from the Soviet Union and its satellites and allies. Three times with large margins the UNGA passed resolutions implicitly condemning Vietnam's continued occupation of Kampuchea. Indeed, a virtual international consensus has been constructed by ASEAN, the United States, and China, ostracizing Vietnam for its aggression. The roll-call votes in the UNGA on these resolutions are shown in Table 5-7.

Probably the most important form of Sino–U.S. cooperation against Vietnam has been a joint underwriting of Thai security. Parallel with Thailand's distancing of itself from the United States, and with the deterioration of Sino–Vietnamese relations, in 1975–1976, was an improvement of China's relations with Thailand. During his visit to China in July, 1975, to normalize Sino–Thai diplomatic relations, Thai Prime Minister Kukrit Pramoj told reporters that Thailand regarded the United States and China as "two equal friends."[83] By early 1976, China was ready to help Thailand solve peacefully any problems it might have with Laos, Cambodia, or Vietnam, was giving Thailand informal and unofficial guarantees regarding its security, and was encouraging Thailand to retain at least a token U.S. military presence.[84] By the time Prime Minister Kriangsak visited China in March–April 1978, China was stressing Thailand's contribution to regional stability and praising ASEAN's efforts to "safeguard national independence and state sovereignty and to oppose hegemonism."[85]

After Vietnam's occupation of Kampuchea brought Vietnam's army to the Thai–Kampuchean border, China and the United States acted jointly to guarantee Thailand against Vietnamese attack. This joint guarantee was made explicit in January 1980 by Secretary of Defense Harold Brown, when he said that if Vietnam attacked Thailand, the United States would welcome Chinese military assistance against Vietnam in whatever form China might choose to render it.[86] This joint security guarantee was illustrated in June 1980 during the brief Vietnamese attack against Khmer refugee camps inside Thailand, when the United States called the Vietnamese move a threat to peace and stepped up its arms deliveries to Bangkok, while China warned Vietnam that it ran "grave danger" if it persisted in its "military adventure" against Thailand.[87]

It is likely that a Sino–U.S. understanding with regard to Thai security was reached in late 1978 or early 1979. At the end of 1978, all parties anticipated a Vietnamese move into Kampuchea and knew that Kampuchean resistance to such an invasion would be effective only if assisted by outside powers. Thailand provided the only contiguous land border with Kampuchea over which such assistance might flow (the others being Kampuchea's borders with Laos and Vietnam), but such supply of Khmer resistance forces through Thai territory would certainly tempt Vietnamese "incursions" to interdict or disrupt supply lines, depots, or staging areas. There were reports from Chinese, Kampuchean, and Hungarian sources that during his trip to Bangkok in November 1978 Deng Xiaoping secured permission for China to supply weapons to Kampuchean

TABLE 5-7. Roll Call Votes on U.N. General Assembly Resolutions Implicitly Condemning Vietnam's Occupation of Kampuchea

No. 34/22 *14 Nov. 1979* *91 for; 21 against*	*No. 35/6* *22 Oct. 1980* *97 for; 23 against*	*No. 36/5* *21 Oct. 1981* *100 for; 25 against*
States voting in favour of resolution 34/22 were: Argentina, Australia, Austria, Bahamas, Bangladesh, Barbados, Belgium, Bhutan, Bolivia, Botswana, Brazil, Burma, Canada, Chad, Chile, China, Colombia, Comoros, Costa Rica, Democratic Kampuchea, Denmark, Djibouti, Ecuador, Egypt, El Salvador, Equatorial Guinea, Fiji, France, Gabon, Gambia, Germany, Federal Republic of, Ghana, Greece, Guatemala, Haiti, Honduras, Iceland, Indonesia, Ireland, Israel, Italy, Japan, Kenya, Lesotho, Liberia, Luxembourg, Malawi, Malaysia, Maldives, Malta, Mauritania, Mauritius, Morocco, Nepal, Netherlands, New Zealand, Niger, Nigeria, Norway, Oman, Pakistan, Papua New Guinea, Paraguay, Peru, Philippines, Portugal, Rwanda, Samoa, Saudi Arabia, Senegal, Singapore, Solomon Islands, Somalia, Spain, Sri Lanka, Sudan, Suriname, Swaziland,	States voting in favour of the resolution were: Argentina, Australia, Austria, Bahamas, Bahrain, Bangladesh, Barbados, Belgium, Bhutan, Bolivia, Botswana, Brazil, Burma, Burundi, Canada, Central African Republic, Chile, China, Colombia, Comoros, Costa Rica, Democratic Kampuchea, Denmark, Djibouti, Ecuador, Egypt, El Salvador, Equatorial Guinea, Fiji, France, Gabon, Gambia, Germany, Federal Republic of, Ghana, Greece, Guatemala, Haiti, Honduras, Iceland, Indonesia, Ireland, Israel, Italy, Japan, Kenya, Kuwait, Lesotho, Liberia, Luxembourg, Malaysia, Maldives, Malta, Mauritania, Mauritius, Morocco, Nepal, Netherlands, New Zealand, Niger, Nigeria, Norway, Oman, Pakistan, Papua New Guinea, Paraguay, Peru, Philippines, Portugal, Qatar, Rwanda, Saint Lucia, Samoa, Saudi Arabia, Senegal, Singapore, Solomon Islands,	States voting in favour of the resolution were: Argentina, Australia, Austria, Bahamas, Bahrain, Bangladesh, Belgium, Belize, Bhutan, Bolivia, Botswana, Brazil, Burma, Burundi, Canada, Central African Republic, Chile, China, Colombia, Comoros, Costa Rica, Democratic Kampuchea, Denmark, Djibouti, Dominican Republic, Ecuador, Egypt, El Salvador, Equatorial Guinea, Fiji, France, Gabon, Gambia, Germany, Federal Republic of, Ghana, Greece, Guatemala, Haiti, Honduras, Iceland, Indonesia, Ireland, Israel, Italy, Jamaica, Japan, Kenya, Kuwait, Lesotho, Liberia, Luxembourg, Malaysia, Maldives, Malta, Mauritania, Mauritius, Morocco, Nepal, Netherlands, New Zealand, Niger, Nigeria, Norway, Oman, Pakistan, Papua New Guinea, Paraguay, Peru, Philippines, Portugal, Qatar, Rwanda, Saint Lucia, Saint Vincent and the

Sweden, Thailand, Togo, Tunisia, Turkey, United Kingdom, United Republic of Cameroon, United States, Upper Volta, Uruguay, Venezuela, Yugoslavia, and Zaire.

States voting against were: Afghanistan, Angola, Bulgaria, Byelorussian SSR, Cuba, Czechoslovakia, Democratic Yemen, Ethiopia, German Democratic Republic, Grenada, Guyana, Hungary, Lao People's Democratic Republic, Mongolia, Mozambique, Nicaragua, Poland, Sao Tome and Principe, Ukrainian SSR, USSR and Viet Nam.

States abstaining were: Algeria, Bahrain, Benin, Burundi, Cape Verde, Central African Republic, Congo, Dominican Republic, Finland, Guinea, Guinea-Bissau, India, Ivory Coast, Jamaica, Jordan, Kuwait, Lebanon, Madagascar, Mali, Mexico, Panama, Qatar, Sierra Leone, Syrian Arab Republic, Trinidad and Tobago, Uganda, United Arab Emirates, United Republic of Tanzania and Zambia.

Somalia, Spain, Sri Lanka, Sudan, Suriname, Swaziland, Sweden, Thailand, Togo, Trinidad and Tobago, Tunisia, Turkey, United Arab Emirates, United Kingdom, United Republic of Cameroon, United States, Upper Volta, Uruguay, Venezuela, Yugoslavia and Zaire.

States voting against were: Afghanistan, Angola, Bulgaria, Byelorussian SSR, Congo, Cuba, Czechoslovakia, Democratic Yemen, Ethiopia, German Democratic Republic, Grenada, Guyana, Hungary, Lao People's Democratic Republic, Libyan Arab Jamahiriya, Mongolia, Mozambique, Poland, Seychelles, Syrian Arab Republic, Ukrainian SSR, USSR, and Viet Nam.

States abstaining were: Algeria, Benin, Cape Verde, Chad, Finland, Guinea, Guinea-Bissau, India, Ivory Coast, Jamaica, Jordan, Lebanon, Madagascar, Malawi, Mali, Panama, Sao Tome and Principe, Sierra Leone, Uganda, United Republic of Tanzania, Yemen and Zambia.

Grenadines, Samoa, Saudi Arabia, Senegal, Singapore, Solomon Islands, Somalia, Spain, Sri Lanka, Sudan, Suriname, Swaziland, Sweden, Thailand, Togo, Tunisia, Turkey, United Arab Emirates, United Kingdom, United Republic of Cameroon, United States, Upper Volta, Uruguay, Venezuela, Yugoslavia, Zaire, and Zimbabwe.

States voting against were: Afghanistan, Angola, Bulgaria, Byelorussian SSR, Chad, Congo, Cuba, Czechoslovakia, Democratic Yemen, Ethiopia, German Democratic Republic, Grenada, Hungary, Lao People's Democratic Republic, Libyan Arab Jamahiriya, Mongolia, Mozambique, Nicaragua, Poland, Seychelles, Syrian Arab Republic, Ukrainian SSR, USSR, Vanuatu, Viet Nam.

States abstaining were: Algeria, Benin, Cape Verde, Finland, Guinea, Guinea-Bissau, India, Lebanon, Madagascar, Malawi, Mali, Mexico, Panama, Sao Tome and Principe, Sierra Leone, Trinidad and Tobago, Uganda, United Republic of Tanzania, Zambia.

Source: "Call for Immediate Withdrawal of All Foreign Troops from Kampuchea," *UN Chronicle,* Jan. 1980, Vol. XVII, No. 1, p. 39. "Assembly Calls for International Conference on Kampuchean Issue," *UN Chronicle,* Dec. 1980, Vol. XVII, No. 10, p. 13. "Assembly Again Stresses Need for Foreign Troops to Quit Kampuchea; Urges that Conference be Reconvened," *UN Chronicle,* Dec. 1981, Vol. XVIII, No. 11, p. 5.

forces via Thai territory.[88] At that time, China expected the Pol Pot regime to hold out longer than it did, and the rapid occupation of all of Kampuchea by Vietnamese forces probably caused the Thai government to reconsider its role in aiding the Khmer Rouge. According to Richard Nations, three factors persuaded Thailand to continue allowing its territory to be used by the anti-Vietnamese Khmer resistance. These factors were: (1) the Khmer Rouge decision to fight within Kampuchea and not flee to Thailand, (2) the immediate ASEAN support for Thailand after Vietnam's move, and (3) China's willingness to act strongly as indicated by its punitive invasion of Vietnam.[39] Assurances of U.S. support were also important. During his visit to Washington in February 1979 (just before China struck Vietnam) it is highly probable that Kriangsak discussed the relation between Thailand's role in assisting the Khmer Rouge and U.S. guarantees of Thai security. In other words, China and the United States agreed to guarantee Thailand against Vietnamese attack so that Thailand was willing to allow its territory to be used to support the anti-Vietnam resistance in Kampuchea.

The incoming Reagan Administration reportedly was "startled by the depth and breadth" of U.S. relations with China as decribed to them in briefings by officials of the outgoing Carter Administration.[90] Once in office, however, the new Administration decided not only to uphold established programs of Sino–U.S. strategic cooperation, but to deepen such cooperation — a move in line with its determination to more forcefully counter Soviet advances in Asia, Africa, and Latin America. In the early spring of 1981, a 30-page CIA paper reportedly was submitted to the National Security Council assessing Chinese interests in Africa and evaluating proposals for U.S.–Chinese cooperation in Africa to thwart Soviet advances.[91] Shortly before Secretary Haig left for China in mid-June 1981, the Administration decided to lift the ban on U.S. arms sales to China, and to consider further Chinese requests for weapons purchases on a case by case basis.[92] This was a highly significant indication that the new Administration hoped to advance the U.S.'s strategic relationship with China.

During their flight to China, officials of the Haig party told reporters that the United States was prepared to discuss strategic and military cooperation with the Chinese in considerable detail if the Chinese were so inclined.[93] Cooperation in Indochina was a major component of the U.S. proposals for expanded cooperation. Haig told reporters aboard his China-bound flight that he intended to stress to the Chinese the convergence of U.S. and Chinese strategic views on Vietnam and the Soviet Union. He condemned Vietnam as a Soviet "proxy" and as the "Cubans" of Southeast Asia, and characterized its activities as constituting "indirect Soviet pressure" against China. The Reagan Administration was said to be "increasingly concerned" about the Vietnamese invasion of Kampuchea and the Soviet presence at Cam Ranh Bay, and hoped that China and ASEAN would take the lead in opposing and containing Vietnam. The United States itself hoped to deepen and continue Vietnam's isolation. It also had a "firm commitment" to advancing the Sino–American strategic relationship.[94]

During Haig's discussions with Chinese officials in Peking, much attention was paid to the Indochinese situation. As Haig told a news conference on June 16:

> Naturally, much of our discussion focused upon the challenges posed by the Soviet Union and its proxies in Afghanistan and in Indochina. We showed a common determination to prevent the pressure of other events from deflecting attention away from this strategic challenge. Our objectives in both areas coincide – above all, in our resolve to press for the complete withdrawal of foreign military forces from Afghanistan and Kampuchea.[95]

The exact contours of Sino–American understandings with regard to coordination of their Indochinese policies can only be surmised. The U.S. side probably agreed to use its influence to cut off various forms of economic assistance still reaching Vietnam. It may have explained to the Chinese the constraints on its supplying of arms to the Kampuchean "third force" (led by Son Sann), but expressed its desire to see this group strengthened. Later reports from Washington indicated that an agreement was reached whereby China would supply arms to various Khmer resistance groups, while the U.S. would ensure that the food distribution network along the Thai–Kampuchean border continued to support the resistance groups.[96] Both sides probably agreed to support ASEAN's plan for a United Nations conference on Kampuchea and for the creation of a front uniting the three major factions of the Kampuchean resistance. It was probably agreed that international attention should be kept focused on the Vietnamese occupation of Kampuchea and that propaganda activities to this effect would be increased. Finally, there may also have been agreement on cooperation in covert operations designed to destabilize Hanoi's rule over Laos, Kampuchea, and Vietnam.

INCREASING THE PRESSURE ON VIETNAM

Vietnam's invasion and occupation of Kampuchea are perceived by U.S. decision makers as a case of imperialist aggression, which, if allowed to succeed, will encourage Hanoi to undertake further aggression against its neighbors. Moreover, its occupation of Kampuchea puts Hanoi in a strong position to destabilize Thailand, if it chooses to do so. Thus the attempt to bring about a withdrawal of Vietnamese troops from Kampuchea and the creation of an independent, neutral Kampuchea has been the crux of U.S. policy in Indochina since Vietnam's invasion of Kampuchea in December 1978. The way in which the United States proposed to do this is by imposing sanctions against Vietnam until it agrees to withdraw its troops. In the words of Assistant Secretary of State Holdrige:

> The central issue in U.S. policy toward Vietnam is the occupation of Kampuchea, and that is why we will continue to keep pressure on Hanoi. In this we and ASEAN are in full agreement: the course of action most likely to result in the removal of Vietnamese troops from Kampuchea is to make the occupation as costly as possible for Hanoi. We will continue a process of diplomatic isolation and economic pressure until Hanoi is prepared to follow the will of the world as expressed in two consecutive U.N. General Assembly resolutions and agrees to troop withdrawal, free elections, and an end to outside interference in Kampuchea.[97]

According to Secretary Haig's speech to the ASEAN foreign ministers' conference in June 1981, the United States has two major goals with regard to Kampuchea: (1) the restoration of a sovereign Kampuchea, free of foreign intervention, and whose government "genuinely represents the wishes of the Khmer people," and (2) the creation of a neutral Kampuchea that represents no threat to *any* of its neighbors.[98]

Reagan's policy of pressuring Vietnam differs from Carter's to the extent that the new Administration has sought additional ways of tightening the vise on Vietnam. The Reagan Administration's effort to punish Vietnam is much more comprehensive and systematic than similar efforts by the Carter Administration. A number of techniques have been used to raise political, economic, and possibly military pressure on Vietnam. In April 1981 the United States supported the ASEAN proposal for a U.N.-sponsored international conference to deliberate on a plan to bring about a realization of earlier UNGA resolutions calling for the withdrawal of Vietnamese troops from Kampuchea.[99] When U.N. Secretary General Kurt Waldheim was hesitant to convene such a conference for fear of alienating the Soviet Union and its allies, Vice-President Bush, Secretary Haig, and President Reagan each met with Waldheim and urged him to take the initiative in organizing the conference.[100] Once the U.N. conference was scheduled, Washington gave it strong support and helped to insure wide participation (i.e., 79 nations attended). As Secretary Haig told the ASEAN foreign ministers on June 20, 1981:

The United States also strongly endorses the convening of an international conference to deal with the Kampuchean issue. I intend to personally participate. We urge all parties, including Vietnam, to join the dialogue which can bring general progress to Southeast Asia.[101]

Although there was a small possibility that Vietnam would decide to participate in this conference (as ASEAN hoped), from the U.S. perspective, its primary function was to focus world attention on Vietnam's occupation of Kampuchea and morally ostracize and politically isolate Vietnam for its aggression.

The Reagan Administration has also attempted to increase the condemnation of Vietnam for human rights violations. Since 1979 the State Department's annual human rights reports had included criticism of communist countries as well as of recipients of U.S. aid,[102] but under Reagan such reports have toned down criticism of Washington's noncommunist authoritarian allies, as well as of China, while focusing on the Soviet Union and its friends and allies.[103] Other forums have been used to issue such propaganda. In a speech on "Human Rights and American Interests" to the Trilateral Commission on March 31, 1981, for example, Secretary Haig implicitly condemned the "totalitarianism" of such countries as the Soviet Union, Cuba, Vietnam, Kampuchea, and Afghanistan.[104]

Probably the Reagan Administration's most important political assault against Vietnam has been the campaign to convince public opinion that Hanoi is using biological weapons against resistance forces in Kampuchea and Laos. (By treating such charges as "propaganda" the author does not mean to imply that the charges are not, quite probably, true.) Speaking in West Berlin on September 13, 1981, Secretary Haig charged that physical evidence from Southeast Asia had been analyzed and found to contain three deadly mycotoxins derived from a fungus and sprayed in liquid form over areas in which anti-Vietnamese guerrillas were operating.[105] At a press conference on September 14, Under Secretary of State Stoessel elaborated on Haig's charges and urged the U.N.'s expert group on biological warfare to send investigators into the affected areas of Indochina.[106] When the December 1981 report of the U.N. expert group proved inconclusive, the United States successfully lobbied to extend for another year the group's mandate to investigate. Then on March 22, 1982, Secretary Haig presented to Congress a detailed report on chemical warfare in Southeast Asia and Afghanistan, which was clearly intended to overcome some of the shortcomings of the report of the U.N. expert group and staunch the media's increasing tendency to discount the U.S's charges as unfounded. Haig's report was later issued as a Special Report by the Department of State and given extensive publicity.[107]

The Reagan Administration has also sought to tighten the economic blockade of Vietnam by cutting off remaining international aid still reaching Vietnam. As Haig told the Manila ASEAN conference: "We will continue to question seriously any economic assistance to Vietnam—whatever the source—so long as Vietnam continues to squander its scarce resources for aggressive purposes."[108] Early in May, 1981, the State Department attempted for the first time since 1975 to block humanitarian food shipments to Vietnam, when it denied a license to the Mennonite Church to export 250 tons of flour to Vietnam.[109] An intensive lobbying effort reversed this particular decision, but it was indicative of the new hardness in U.S. policy toward Vietnam.[110]

More important than attempts to block private philanthropy is Washington's new determination to cut multilateral assistance to Vietnam and to Vietnamese-occupied Kampuchea. In May 1981 U.S. representatives in U.N. agencies were instructed to argue against any aid to Vietnam on the grounds that such aid amounted to a subsidizing of aggression condemned by successive U.N. General Assembly resolutions.[111] In June 1981 the U.N. Development Program Council met to consider assistance on the order of $94 to $118 million over the 1982–1984 period. Although the U.S. representatives argued against such assistance, the 46-nation council nevertheless approved $18 million in assistance to Vietnam for fiscal year 1982 and $118 million over the full period.[112] The United States also tried to block the approval of a $5-million Food and Agriculture Organization program to Vietnam.[113] Washington may also be using its influence to delay shipments of multilateral aid to Vietnam.[114]

An effective form of U.S. pressure would be the blockage of International Monetary Fund (IMF) approval of a Vietnamese request to draw down its secured credit tranche. Vietnam's request to the IMF for its first credit tranche of $28 million was approved by the IMF board in January 1981. Although the U.S. representatives were understood to be unhappy with this decision, since it was the first tranche, the IMF board had little discretion. Second and subsequent tranches involve greater latitude for discretion, however, and if Hanoi were to apply for the second tranche of its $135-million Special Drawing Rights (SDR) quota, the U.S. representatives could express their concern with the irrationality of Vietnam's pricing structure and other strictly economic considerations that would be a basis for IMF denial.

The United States is also seeking to cut international assistance to Kampuchea. As the acting Director of the State Department's Bureau of Refugee Programs told the House Asian and Pacific Affairs Subcommittee; ". . . mindful of concerns about development assistance inside Vietnamese-occupied Kampuchea, we and other donors have pressed for the termination of activities by the U.N. Joint Mission for Cambodia Relief as soon as the Khmer are able to feed themselves, or if shortfalls in food self-sufficiency continue, as soon as they are manageable."[115] The U.N. Kampuchean Relief Program ended in December 1981. The United States urged that until that time, aid be kept as close as possible to levels necessary for humanitarian relief, and not assume the proportions of developmental aid.[116]

A few months after assuming office, the Reagan Administration decided to actively support an ASEAN effort, begun in March 1981,[117] to create a united front of all major Khmer groups opposed to the Vietnamese occupation. In order to encourage Khmer unity it was decided to give a higher profile to U.S. contacts with Son Sann (the leader of the noncommunist "third force"), and to encourage him to play a prominent role in the Kampuchean resistance.[118] In early March, U.S. Ambassador to Thailand Morton Abramowitz met with Son Sann; and in late April, Son Sann came to the United States for a series of meetings with U.S. officials, including Under Secretary Stoessel and Assistant Secretary Holdridge. It was reported that U.S. officials urged Son Sann to find a way of forming a coalition with the Khmer Rouge forces, while maintaining political superiority within the coalition and insuring continued active Khmer Rouge resistance to the Vietnamese occupation forces.[119] In mid-1981, if Norodom Sihanouk is to be believed, U.S. chargé d'affairs in Peking, J. Stapleton Roy, urged him to form a united front with the Khmer Rouge, saying that this would make it easier for friendly countries to aid him.[120] It seems likely that behind the scenes U.S. influence helped give birth to the tripartite declaration of the major Khmer resistance groups on September 4, 1981[121]

Washington has also given low-key support to the formation of a pan-Indochinese united front of groups opposed to Hanoi's current policies. In July 1981, the U.S. embassy in Paris issued a visa to former minister of justice of the Provisional Revolutionary Government of South Vietnam Truong Nhn Tang, enabling him to come to New York at the time of the U.N. conference on Kampuchea.[122] Shortly before coming to the United States, Tang had set up an organization committed to the overthrow of the Hanoi government. While in Washington, Tang met with officials from the State Department as well as Senator John Tower, chairman of the Senate Armed Services Committee, and laid out his proposal for an anti-Hanoi front of Lao, Khmer, and Vietnamese groups. State Department officials also met with leaders of various Laotian resistance groups.[123]

The United States provides support to the Khmer resistance forces through its cross-border food distribution system in Thailand. Although the "land bridge" openly distributing relief food across the border to needy people in Kampuchea has been discontinued, there is substantial Khmer movement back and forth across the border.[124]

The U.S. government has denied that Washington is supplying financial aid or military equipment to the noncommunist Khmer resistance groups. Assistant Secretary Holdridge, however, did hint at the possibility of aid to the Kampuchean resistance in his speech to the American Club in Peking during Secretary Haig's visit there in June 1981. He said that the United States would "see if we can find ways to increase the political, economic, and, yes, military pressures on Vietnam, working with others in ways which will bring about, we hope, some change in Hanoi's attitude toward the situation."[125] Holdridge's statement was taken as a "trial balloon" by the media, but the reaction was strongly negative. "Would American public opinion stand still for one minute for an indirect re-entry into the Indochinese wars?" asked a *Washington Post* editorial.[126]

When Haig was later asked whether or not the United States would provide military assistance to the Kampuchean resistance forces, he replied that he knew of no decision to do this, nor of any suggestion of a decision to do this.[127] A cynic might interpret Secretary Haig's response as an indication that he had chosen not to be informed of any decision to arm the Khmer resistance groups. Speaking on National Public Radio on August 18, 1981, Under Secretary of State James Buckley indicated the United States had sympathy for the genuine "freedom fighters" of Kampuchea and Afghanistan and stated that the United States had not "discounted" the possibility of military assistance to those freedom fighters.[128] Nevertheless, in spite of much suggestive utterances, there is no evidence that the United States has assumed the role of aiding the Khmer resistance forces.

There is also the possibility that the Reagan Administration has authorized covert operations designed to destabilize Hanoi's rule over Indochina. One assumes that the CIA is involved in intelligence collection operations throughout Indochina. The question of "special operations" is another matter. With regard to the latter the evidence is only circumstantial. During a 2-hour meeting on June 15, 1981, for example, aside from Haig and Chinese Defense Minister Geng Biao, former Deputy Director of the CIA Vernon Walters and Vice-Chief of Chinese Military Intelligence Zhang Zhuzi were also in attendance.[129] Also suggestive were recent comments by Vang Pao, leader of the Hmong mercenary army in Laos, organized, financed, and directed by the CIA in the 1960s. His men, Vang said, had been helping the United States find evidence of Vietnamese use of chemical and biological weapons in Laos.[130] It has also been charged that the U.S.–Kampuchean Emergency Group operating out of Aranyaprathet on the Thai–Kampuchean border to provide relief to Khmer refugees provides cover for an extensive CIA operation designed to destabilize the Kampuchean economy.[131] Unless one sees the U.S. distribution of relief aid in this light, however, there is no evidence to substantiate this charge.

Counterfeit Kampuchean currency is apparently being smuggled into Kampuchea from Thailand.[132] Again, there is no evidence linking this phenomenon to the CIA. Indeed, the poor quality of this counterfeit currency would seem to exonerate the CIA.

It should be noted, however, that the CIA very probably did build up a "stay behind" apparatus in South Vietnam prior to 1975, that China probably does have a covert capability in northern Vietnam and among the overseas Chinese communities of Indochina, and that many Vietnamese, Laotians, and Khmers have been alienated by Hanoi's post-1975 policies. On the other hand, the Reagan Administration has been quite forthright about its intent to reinvigorate the CIA.

U.S. MANAGEMENT OF CHINA–ASEAN CONTRADICTIONS

Washington's effort to build a strategic consensus in Southeast Asia based on the roll back of Vietnamese expansionism faces a number of difficulties. Perhaps the major problem confronting this policy is managing the contradictions between the policies and objectives of China and the ASEAN countries. The United States must persuade those ASEAN countries skeptical of China's long-range objectives, most particularly Indonesia and Malaysia, that Vietnam in fact poses a greater threat to them than does China. If it is unsuccessful in this, ASEAN may drift apart with Indonesia and Malaysia reaching an accommodation with Vietnam for the sake of containing China. On the other hand, it is possible that because of its zeal for containing the Soviet Union and Vietnam, the United States may ignore legitimate and realistic ASEAN concerns and help lay the basis for further Chinese predominance in the whole Southeast Asian region. We shall return to the latter possibility in the concluding section.

China's long-range objectives in Southeast Asia may well conflict with those sought by the ASEAN countries. China's general strategy vis-a-vis Vietnam is to mobilize maximum military, political, and economic pressure against Hanoi, thereby weakening it to the point where it is ready to withdraw its troops from Kampuchea, break with the Soviet Union, and conduct itself in a fashion not fundamentally antagonistic to China's interests. Peking seeks a Vietnam which will be more sensitive to China's wishes and interests on a whole range of issues, from the Soviet presence in Indochina, to the disputed sea floor and archipelagos of the South China Sea, to the ethnic Chinese minority within Vietnam. It is also determined to block Hanoi's efforts to foster a "special relationship" (in Hanoi's words) or "Indochinese federation" (in Peking's words) among the three Indochinese countries. It apparently believes that its interests will be best served by a continued division of Indochina into three less powerful states. In Kampuchea, Peking apparently desires the return of the Democratic Kampuchean Khmer Rouge regime to power in Phnom Penh and opposes anything that undermines the claim of Democratic Kampuchea to the title of legitimate government of Kampuchea or that would block its "postliberation" return to power. China supplies liberal amounts of arms and aid to the Democratic Kampuchean forces while providing much smaller allotments to the noncommunist resistance groups, thereby insuring that the Khmer Rouge remain the dominant anti-Vietnamese military force in Kampuchea.

The perspectives and objectives of the ASEAN countries differ significantly from those of China. While they seek a Vietnamese withdrawal from Kampuchea, they do not desire a Khmer Rouge return to power. They are not persuaded that the latter's repentance for their former barbarities would preclude a repetition of them once the Khmer Rouge leadership was again ensconced in Phnom Penh. Moreover, rather than seeking the disintegration of the current Heng Samrim regime under a military onslaught from insurgent forces, they seek a political settlement that would include a role for the incumbent regime.

The ASEAN governments are apprehensive about the consequences of Southeast Asia becoming an arena for the intense Sino-Soviet rivalry. They tend to perceive that the success of a program designed to force Vietnam into submission would greatly increase Chinese influence in Indochina and throughout Southeast Asia. The failure of such a program, on the other hand, would have the consequence of increasing the Soviet role and presence in Indochina. There are, of course, differences in perspective between Indonesia and Malaysia on the one hand and Thailand and Singapore on the other, with regard to the danger of expanded Chinese influence in Southeast Asia. Nevertheless, the ASEAN countries are generally agreed that every effort should be made to provide Hanoi with a reasonable and early exit from Kampuchea and from its international isolation, thereby leaving Vietnam a strong, prosperous, and independent nation and minimizing both Chinese and Soviet influence in Southeast Asia.

A primary task of American diplomacy is to "correctly handle" these contradictions between China and ASEAN in a manner that will not undermine the strategic consensus/united front against the Soviet Union and its Vietnamese allies. This is, however, a midrange strategy and leaves unanswered the question of long-range U.S. objectives in Indochina and Southeast Asia—a question which is addressed in the concluding section.

One principal aspect of U.S. management of these contradictions has been to rally Indonesian and Malaysian support for the more hard-line position of Thailand. (Geography dictates that Thailand, more than other

ASEAN states, be concerned with reestablishing Kampuchea as a neutral, independent "buffer" between itself and Vietnam. It is also more dependent on Chinese deterrent support against Vietnam.) When Deputy Assistant Secretary of State for East Asia and the Pacific John Negroponte visited ASEAN capitals in March 1981, for example, he urged Indonesian officials to be sensitive to Thailand's distinctive front-line situation.[133] Negroponte also assured ASEAN leaders that the United States felt it was important to retain the Khmer Rouge's U.N. seat,[134] thereby reaffirming U.S. solidarity with China on this issue and indicating that the United States did not intend to follow Australia's lead in derecognizing Democratic Kampuchea.

Many of the contradictions between ASEAN and China came into focus at the July 1981 U.N. Conference on Kampuchea. ASEAN seriously desired that Hanoi and Phnom Penh send a representative—something opposed by China.[135] However the sharpest ASEAN-Chinese disagreements were about measures designed to insure that the postsettlement Kampuchean regime would reflect the genuine and noncoerced sentiments of the Khmer people. Such ASEAN-proposed measures included the dispatch of a U.N. peacekeeping force, the disarming of all Khmer factions, and the establishment of an interim administration in Phnom Penh pending the conduct of internationally supervised elections to establish a new Kampuchean government. China strongly opposed these measures on the grounds that they would constitute interference in the internal affairs of Kampuchea and that the legitimate government of Kampuchea is the Democratic Kampuchea regime. In China's view the Heng Samrin regime was to be excluded from interim consultation on preparations for elections and its forces alone were to be disarmed.[136] The U.S. approach to this dispute was to urge the ASEAN countries to compromise with and not confront China. Moreover, U.S. representatives refused to argue for the ASEAN position.[137]

The United States' own position with regard to the nature of a postsettlement Kampuchean regime seems to be closer to the ASEAN position than to China's. Asked at a news conference in Peking, on June 16, 1981, about U.S.–Chinese agreements on the establishment of an anti-Vietnamese Khmer united front (an issue with clear implications for the nature of the postsettlement Kampuchean regime), Secretary Haig replied:

> I think we have some differences of nuance on that issue, but we are essentially of one mind that a united front or a front that would be representative of the wishes and the aspiration of the Khmer people be formed and that free elections determine the ultimate outcome of the final regime there, and in no way could the current puppet regime established by Hanoi be representative of a popular government.[138]

The two parts of Haig's answer were contradictory. It is understandable that the Secretary of State would want to minimize Sino–U.S. differences, but for our purposes the initial clause is the most important. The "nuances" Haig was referring to probably had to do with the role of the Khmer Rouge. In a prepared testimony before the Asia–Pacific Subcommittee of the Senate Foreign Relations Committee in July 1981, Assistant Secretary Holdridge elaborated the U.S. view of the Khmer Rouge:

> It is intolerable to acquiesce in a situation created by an invading army and perpetuated by a massive occupation by foreign troops. We hold no grief for the Pol Pot regime thrown out of Phnom Penh by the invading Vietnamese. It was abominable in its treatment of the Khmer people, and we can under no circumstances favor its return to power.[139]

Washington does not perceive its support for continued seating of Democratic Kampuchea in the U.N. General Assembly as support for the Pol Pot regime. Rather, it sees expulsion of Democratic Kampuchea from the United Nations as legitimizing the Heng Samrin regime and as lessening the pressure on Vietnam to withdraw from Kampuchea.[140] Washington assumes, along with ASEAN, that if the Khmer people are allowed to freely choose their rulers, those rulers will not include the Khmer Rouge. American support for U.N. supervised, free elections in Kampuchea (as endorsed in the ASEAN-sponsored U.N. General Assembly resolution 35/6) thus translates into a minimal role for the Khmer Rouge in the post-Vietnamese-withdrawal Kampuchea government. In his address to the U.N. conference on Kampuchea, Secretary Haig clearly stated U.S. objectives with regard to the nature of the postsettlement Kampuchean regime:

> The position of the United States is clear: We believe that the world community has an obligation to assure the Khmer people their right to choose their own government and to live in peace and dignity. We, therefore, see this conference as having two closely related goals:
>
> The restoration of a sovereign Kampuchea...whose government genuinely represents the wishes of the Khmer people; and a neutral Kampuchea that represents no threat to any of its neighbors.[141]

Given the performance of the Khmer Rouge when they ruled Kampuchea, a strong case can also be made that the objective of a nonthreatening Kampuchea would preclude renewed Khmer Rouge rule.[142]

In spite of its desire to see a postsettlement Kampuchean government in which the Khmer Rouge will play a minimal role, the United States helped arrange a compromise on this issue at the U.N. conference on Kampuchea, which conceded the essence of China's demands by dropping explicit provisions designed to insure the Khmer people a free choice of rulers.[143]

There are several feasible explanations of U.S. willingness to compromise on this critical issue. One would be that the future welfare of the Khmer people was considered less important than consolidating Sino–American cooperation against the Soviet Union. It seems more likely, however, that rather than making a deliberate choice, U.S. diplomats persuaded themselves that since any settlement in Kampuchea is unlikely in the near future, by the time such a settlement does come about, the noncommunist resistance forces will be stronger than those of the Khmer Rouge, thereby obviating the need for explicit safeguards. It is also possible that the United States is not unhappy with continuing Vietnamese occupation of Kampuchea, since this provides a durable base for an anti-Soviet strategic consensus, and that the United States was, therefore, quite ready to drop measures that would have encouraged Vietnam to compromise. This solution, however, seems too clever to believe. Perhaps most likely is the possibility that the problem was not thought through at all and that the compromise was simply arranged as a pragmatic way of minimizing conflict between Washington's Chinese and ASEAN partners.

CONTINUITIES, DISCONTINUITIES, AND LONG-TERM OBJECTIVES IN U.S. SOUTHEAST ASIAN POLICY

It is now possible to return to the question of continuity and change with which we began this chapter, and to draw some general conclusions about the long-term objectives that current U.S. policy in Southeast Asia is designed to serve. In two key ways current U.S. Southeast Asian policy clearly is similar to the "traditional" U.S. policy in that region: (1) U.S. decision makers believe that Washington has very important economic, strategic, and political interests in the ASEAN region of Southeast Asia, and (2) Indochina is perceived as a stepping stone to the rest of Southeast Asia. In terms of discontinuities, the Reagan and Carter Administrations seem to have abandoned the view that domination of Indochina by the power ruling China poses unacceptable threats to vital U.S. interests in the rest of Southeast Asia. This latter point is the conclusion one reaches from analyzing the probable consequences of U.S. policy toward Indochina. The United States is pursuing policies which will have the long-range consequence, if they are successful, of greatly increasing Chinese influence in Indochina. If Hanoi does indeed "break," expels the Soviet military presence from Vietnam, and agrees to act in a more respectful way toward its great northern neighbor, and if China's Khmer Rouge clients return to power in Phnom Penh, China's standing in Indochina, and throughout the whole of Southeast Asia, would be very substantially enhanced.

How does one explain this apparent departure from traditional U.S. objectives in Southeast Asia? There seem to be four possible explanations for this discontinuity: (1) a deliberate concession of Indochina to China as a sphere of influence; (2) a belief that China does not and will not attempt to dominate Indochina; (3) a plan to produce a nonexpansionist but independent Vietnam, and (4) an un-thought-through policy of muddle.

It is conceivable that a decision could have been made in Washington, in effect, to write off Indochina as ultimately destined to become a Chinese sphere of influence. Such a policy would presumably be based on several solid arguments and would probably involve a recognition of a "new China," which is likely to be a global power by the end of this century, and which will seek, and is entitled to, a sphere of influence in Southeast Asia—just as the Soviet Union dominates Eastern Europe and the United States views Central America and the Caribbean as its "backyard." In the late 1960s, some analysts of U.S. Asian policy advocated that the United States build a long-term, peaceful, and cooperative relationship with China by withdrawing from the continent of Asia and recognizing mainland Southeast Asia as a Chinese sphere of influence.[144] Such a policy would arguably help minimize future tension between the United States and the emerging Chinese state. Conceding a Chinese sphere of influence in Indochina could also be a way of maximizing chances for long-term cooperation between China and the United States in containing the Soviet Union. It would also be based on the real limits of U.S. power in shaping events in this remote region.

A number of considerations, however, lead to the conclusion that such a brave, grand decision has *not* been made in Washington. Would such a decision have remained secret in leak-prone Washington, or would decision makers dare to assume that it would remain secret? Would U.S. leaders be capable of making such a hard decision except in the most extreme circumstances? Is it likely that a Reagan Administration that is intent on restoring U.S. global preeminence would knowingly turn Indochina over to Chinese domination? Finally, since the Reagan Administration does, as we have seen, perceive major U.S. interests at stake in ASEAN–Southeast Asia and views Indochina as a springboard to that region, it is extremely unlikely that it would consciously accept Chinese domination of Indochina.

The second possible explanation of the "untraditional" U.S. policy (i.e., that Washington believes China does not intend to dominate Indochina) also proves on examination to be unsatisfactory. One does not need to doubt the subjectively altruistic nature of China's intentions to conclude that the fulfillment of its objectives would amount to *de facto* Chinese hegemony over Indochina. For Washington to close its eyes to the latter reality because of a faith in China's good intentions would contrast sharply with the Realpolitik approach that has typically underlaid U.S. Asian policy. It is difficult to see why Washington would suddenly place such faith in pledges unbacked by power. A Realist approach would indicate that if Vietnam is in fact brought to its knees, China will take advantage of this to project it influence throughout Indochina and beyond, whatever the coloration of China's subjective intentions. Thus, it is more probable that a professed U.S. faith in China's antihegemonist pledges is a facade for either of the two remaining policies.

The third possible explanation of U.S. objectives—that Washington expects ultimately to produce a nonexpansionist but also a non-Chinese-dominated Vietnam—is more satisfactory than the first two explanations. Indeed, it proves to be the most satisfactory of the rational actor explanations. Even while mobilizing maximum pressure on Vietnam, Washington may expect that the dogmatic and tenaciously nationalistic leadership in Hanoi will never "break" to the extent that they accept Chinese tutelage. U.S. decision makers may believe that at some point Hanoi will agree to a compromise settlement in Kampuchea, permitting the reestablishment of a neutral Kampuchean buffer state between Thailand and Vietnam, but with Vietnam retaining its determination to resist Chinese dictation. At that point, Vietnam could swing into alignment with other Southeast Asian states to resist Chinese advances, and the Western campaign to punish and isolate Vietnam could be abandoned. Thailand's security would thus be maintained, while China would be effectively contained by a combination of pro-West ASEAN states and a Titoist Vietnam. It seems likely that this is ASEAN's view of the most desirable outcome. It is also the most persuasive of the various "rational" explanations of U.S. policy, although it is diplomatically left unstated for the sake of good relations with China.

A final possibility is that the long-term consequences of U.S. policy have not been thought through by Washington, and that U.S. policy is basically an ad hoc response to a collection of emotions, immediate pressures, and short-term objectives. The Vietnam War is the only war that the United States has clearly lost and the men deciding U.S. foreign policy are still, by and large, the ones who decided it in the 1960s. The desire to punish the victors of that war may be a strong motive in the subconscious minds of U.S. decision makers. Moreover, Vietnam's conquest of Kampuchea was "aggression" in the sense that we usually think of that term, and since the experience of the 1930s the reflexive response of Western leaders to aggression has been to punish, contain, or undo such aggression lest further and bolder aggression occur. There are also substantial short-term "strategic" benefits derived from Sino–U.S. cooperation against Vietnam. Such cooperation is one important component of a broader Sino–U.S. strategic entente. Moreover, Indochina has become the major issue in dispute between China and the Soviet Union, and by stressing mutual U.S.–Chinese interests in that conflict Washington minimizes chances for a Sino–Soviet reconciliation.

The problem with these latter two perspectives is that in spite of the apparently "strategic" nature of the considerations involved (e.g., anti-appeasement, Sino–U.S. strategic entente) they are essentially tactical in that they do not address the question of the configuration of power that the United States desires to ultimately see in Southeast Asia. Although such tactical policy considerations do not necessarily contradict one or another desired configuration of power in Southeast Asia, neither are they necessarily linked to a particular configuration. In other words, it may be that the United States is acting in a "firm" and "decisive" fashion in Southeast Asia without really knowing where such actions will lead.

NOTES

1. H. Feis, *The Road to Pearl Harbor*, Anthenum Books, 1965. See especially pp. 72–74, 85–87, 149.

2. J. H. Holdridge, testimony to East Asian and Pacific Affairs Subcommittee of Senate Foreign Relations Committee, in *U.S. Interests in Southeast Asia*, U.S. Department of State Bureau of Pacific Affairs, Current Policy No. 295, July 1981, p. 2.

3. *Ibid.*

4. *Mineral Yearbook, 1980, Volume III, Area Reports: International*, U.S. Department of Interior, U.S. Bureau of Mine, Washington, D.C., 1982, p. 19. (Hereafter cited "Mineral Yearbook, 1980.").

5. *Ibid*, p. 43. International Monetary Fund (IMF), *Direction of Trade Yearbook, 1980–81*, Washington, D.C., 1981, p. 13. Statement by Michael Armacost "FY 1982 Assistance Requests" in *Department of State Bulletin*, May 1981, O1. 81, No. 2050, p. 28.

6. *Foreign Assistance Legislation for Fiscal Year 1982 (Part 5)*, Hearings and markup before the subcommittee on Asia and Pacific Affairs of the Committee on Foreign Affairs, House of Representatives, 97th Congress, 1st session, March 23–31, April 6, 1981, pp. 36–38. (Hereafter cited "Foreign Assistance Hearings 1982.")

7. *Mineral Yearbook, 1980*, pp. 479–480.

8. *Ibid.*, pp. 637–639.

9. *Ibid.* Also "Net U.S. Imports of Selected Minerals and Metals as per cent of apparent consumption, 1960 to 1979, and by Major Foreign Sources, 1975–1978," in *U.S. Statistical Abstract 1980*, U.S. Department of Commerce, Bureau of the Census, 1981, p. 764.

10. *Mineral Yearbook, 1980*, p. 782.

11. *Ibid.* pp. 31, 32, 34.

12. *Ibid.* pp. 959–963.

13. Regarding the degree and nature of Japanese economic dependence on Southeast Asia see, *Japan in Southeast Asia, Collision Course*, Raul S. Manglapus, Washington: Carnegie Endowment for International Peace, 1976.

14. "*Foreign Assistance 1982 Hearings*," p. xv.
15. Michael Richardson, "Missile Maneuvers," *Far Eastern Economic Review* (hereafter cited "FEER"), April 30, 1982, pp. 32–33.
16. *Ibid.*
17. *Ibid.*
18. Holdridge, *U.S. Interests in Southeast Asia*, p. 2.
19. IMF, *International Financial Statistics Yearbook, 1981*, pp. 229, 289, 353, 377, 415.
20. World Bank, *World Bank Annual Report 1981*, p. 43.
21. Holdridge, *U.S. Interests in Southeast Asia*, p. 3.
22. *Ibid.*
23. Richard Burt, "Soviet Ships Arrive at Cam Ranh Bay," *New York Times* (NYT) March 29, 1979. Marjorie Neihaus, "Southeast Asia," in *Chronologies of Major Developments in Selected Areas of Foreign Affairs*, Committee on Foreign Affairs, U.S. House of Representatives, Committee Print Cumulative Edition 1980, p. 219.
24. Prepared statement by Michael A. Armacost, Deputy Assistant Secretary of State, Bureau of East Asian and Pacific Affairs, in *Foreign Assistance Hearings, 1982*, p. 36.
25. See for example, John McBeth, "The Bulldozer Invasion," *FEER*, May 8, 1981, pp. 26–28; "Hazards along the neutral path," *FEER*, September 19, 1980, pp. 43–46; "A frontier of fear and factions, *FEER*, June 20, 1980, pp. 16–19.
26. Regarding the strength of the various Philippine insurgencies, see, Sheilah Ocampo, "Guns are not the only answer," *FEER*, May 8, 1981, pp. 40–42. Regarding the broader socio-economic problems of the Philippines, see the articles in the same journal over the past several years.
27. *Facts on File, 1979*, p. 537.
28. Holdridge, *U.S. Interests in Southeast Asia*, p. 1.
29. See *Facts on File, 1975*, pp. 188, 429–430, 622; *Facts on File, 1976*, pp. 223, 255.
30. Ho Kwon Ping and Cheah Cheny Hye, "The military shopping list grows longer," *FEER*, October 24, 1980, p. 37.
31. See *Facts on File, 1977*, p. 72.
32. See *Facts on File, 1976*, p. 148; *Facts on File, 1977*, p. 514.
33. Richard Burt, "U.S. is formulating new policy on Asia," *NYT*, March 15, 1979, p. A7.
34. Holdridge, *U.S. Interests in Southeast Asia*, p. 2.
35. *Ibid*, p. 1.
36. Richard Nations, talk on the Thailand–China–Vietnam triangle at the Center of Sino–Soviet Studies, George Washington University, March 16, 1981. (Nations was the Bangkok correspondent for the *Far Eastern Economic Review* throughout the late 1970s.)
37. Holdridge, *U.S. Interests in Southeast Asia*, p. 3.
38. While participating in ceremonies in Australia on May 1, 1982 commemorating the 30th anniversary of the ANZUS treaty, Vice President Bush noted that the treaty remained the "Cornerstone of (U.S.) security policy in the Southwest Pacific," *Facts on File, 1982* p. 340. During Japanese Prime Minister Ohira's visit to Australia in early 1980, Japanese and Australian officials discussed the idea of a pacific community, including Japan, Australia, New Zealand, the U.S.A., Canada and ASEAN as well as the smaller Pacific island nations.
39. Ho Kwon Ping and Cheah Cheng Hye, "Five fingers on the trigger," *FEER*, October 24, 1980, p. 33. Also, *Facts on File, 1976*, p. 148.
40. In 1977 Malaysia and Thailand began cooperating in suppressing the communist-led insurgencies in their common border areas on the Kra Ismus. By the early 1980s the ASEAN countries regularly shared military intelligence and information, conducted officer training exchanges, had begun discussing arms standardization, and periodically carried out joint military maneuvers. See Ho and Cheah, *Ibid.*

41. For example, in November 1980 the Philippines dropped its claim to the Malaysian state of Sabah. *Facts on File, 1980*, p. 902.
42. *Facts on File, 1978*, p. 360.
43. *Facts on File, 1977*, pp. 203, 904.
44. *Facts on File, 1978*, p. 360.
45. Richard Nations, talk at George Washington University, op. cit.
46. Office of the Historian, Bureau of Public Affairs, U.S. Department of State, *Lists of Visits of Foreign Chiefs of State and Heads of Government to the United States, 1789–1978*, research project no. 495 A, 12th Revision, Jan. 1979, p. 84. (Hereafter cited "Lists of Visits")
47. *Facts on File, 1979*, p. 84.
48. See, John McBeth, "Forewarned and forearmed," *FEER*, October 3, 1980, pp. 20–21. Between July 30, and September 14, 1980 at least 10 warships visited Thai Ports.
49. Alexander Haig, "Arrival statement, Manila, June 17, 1981," in *Department of State Bulletin*, August 19, 1981, Vol. 81, No. 2053, p. 40.
50. *Facts on File, 1982*, p. 340.
51. *Facts on File, 1977*, p. 83.
52. *Facts on File, 1978*, p. 36n.
53. *Facts on File, 1981*, p. 505.
54. *Facts on File, 1977*, p. 824.
55. *Facts on File, 1978*, p. 1000, *Facts on File, 1982*, p. 256.
56. Sheilah Ocampo, "Curbing a competitor," *FEER*, April 9, 1982, p. 10.
57. *Ibid*.
58. Helen Ester, "A new test of friendship," *FEER*, February 13, 1981, p. 33.
59. *Facts on File, 1981*, pp. 20, 162, 198.
60. *Facts on File, 1981*, pp. 52, 91.
61. Michael Richardson, "Missile Manoeuvres," *FEER*, April 30, 1982, pp. 32–33.
62. Ho and Cheah, "Five fingers on the trigger," p. 36.
63. Michael Richardson, "Give us the tools," *FEER*, April 2, 1982, p. 30.
64. John Lewis, "Watchdog in the east," *FEER*, May 15, 1981, pp. 11–12.
65. For example see, James Bartholmew, "Gentle persuasion," *FEER*, April 2, 1982, pp. 31–32. Richard Hallaran, "U.S. Presses Japan on Military Role," *New York Times*, December 22, 1981. Henry Scott Stokes, "Japan Seems to Take on Bigger Defense Role," *New York Times*, January 4, 1982. Henry Scott Stokes, "Debate on Defense is Widened in Japan," *New York Times*, April 9, 1981.
66. Ho and Cheah, "Five fingers on the trigger," pp. 32–35.
67. *Washington Post*, June 23, 1981, p. A8.
68. Ho and Cheah, "Five fingers on the trigger," p. 32.
69. Harold Brown, *Annual Report, Fiscal Year 1982*, U.S. Department of Defense, p. 34.
70. *"Foreign Assistance Hearings," 1982*, p. xviii.
71. *Facts on File, 1979*, p. 84.
72. *Facts on File, 1980*, p. 68.
73. Niehaus, op. cit., p. 220. *Facts on File, 1980*, p. 491.
74. *Foreign Assistance Hearings, 1982*, pp. 37–38, 127–128.
75. *Ibid*, pp. xv–xvi.
76. Derek Davies, "The U.S. and the refugees: From charity to cynicism," *FEER*, July 17, 1981, pp. 29–30.
77. "Khmer Relief Efforts," *U.S. Department of State Bulletin*, July 1981, p. 23.
78. Richard Nations, "Stay huddled masses," *FEER*, October 2, 1981, pp. 18–19.
79. *Ibid*.
80. Regarding the disagreements between Brzezinski and Vance see: "Tug Of War over Foreign Policy," *U.S. News and World Report*, June 19, 1978, pp. 37–40; "America's Split Personality," *FEER*, June 6–12, 1980, pp. 30–34. Carter's decision to move ahead with

Sino–U.S. normalization after Brzezinski's return from China in June was mentioned in a talk by Michiel Oksenberg at the Center for Chinese Studies of the University of Michigan, November 2, 1979.

81. Gareth Porter, "The 'China Card' and U.S. Indochina Policy," *Indochina Issues*, Nov. 1980, No. 11, Washington: Center for International Policy.

82. Dan Tretiak, "China's Vietnam War and Its Consequences," *China Quarterly* December 1971, No. 80, pp. 740–767.

83. "Quarterly Chronicle," *China Quarterly*, December 1975, No. 64, p. 811.

84. "Quarterly Chronicle," *China Quarterly*, June 1976, No. 66, p. 447.

85. "Quarterly Chronicle," *China Quarterly*, June 1978, No. 74, p. 478.

86. *New York Times*, January 17, 1980, p. A3.

87. Ron McCrea, "China Warns Vietnam to Halt Thailand Raids," *Washington Post*, June 26, 1980, p. 1.

88. "Quarterly Chronicle," *China Quarterly*, June 1971, No. 78, p. 429. Also "Quarterly Chronicle," *China Quarterly*, March 1979, No. 77, p. 196. These reports were denied by the Thai government, but such denials would be most useful if the reports were true.

89. Richard Nations, talk at George Washington University, March 16, 1981.

90. *New York Times*, December 8, 1980, p. A1, A7. Among the secret Sino–American agreements reviewed and upheld by the new Administration in the spring of 1981 were those providing for joint electronic monitoring of Soviet activities from posts inside China. *Washington Post*, June 18, 1981, p. A3, A4.

91. *Foreign Report*, May 7, 1981, p. 3.

92. *Washington Post*, June 27, 1981, p. A3.

93. *Washington Post*, June 16, 1981, p. A3.

94. Don Oberdorder, *Washington Post*, June 14, 1981, p. 1. See also *New York Times*, June 14, 1981, pp. 1, 24.

95. *Department of State Bulletin*, August 1981, Vol. 81, No. 2053, p. 35.

96. Nayan Chanda, "Ganging up with exiles," *FEER*, July 31, 1981, p. 11.

97. Holdridge, *U.S. Interests in Southeast Asia*, p. 3.

98. "Secretary Haig, July 13, 1981." See *Ibid*.

99. See Address by Undersecretary of State for Political Affairs Walter J. Stoessel to the Los Angeles World Affairs Council, April 24, 1981, in *Department of State Bulletin*, June 1981, Vol. 81, No. 2051, p. 35.

100. *Washington Post*, April 28, 1981, p. 1.

101. *Department of State Bulletin*, August 1981, p. 40.

102. See, "State Department Report on Human Rights," in *Historical Documents of 1980*, Congressional Quarterly, pp. 189–198.

103. *New York Times*, February 10, 1981, p. 4.

104. *New York Times*, April 20, 1981, p. 1.

105. *Los Angeles Times*, September 14, 1981, pp. 1, 6. *Los Angeles Times*, September 15, 1981, p. 1, 6.

106. Richard Nations, "Storm clouds over rain" *FEER*, September 18, 1981, p. 14.

107. Special Report no. 98, U.S. Department of State, *Chemical Warfare in Southeast Asia and Afghanistan*, Report to the Congress from Secretary of State Alexander M. Haig, Jr., March 22, 1982. This propaganda campaign relating to Vietnamese-Soviet chemical warfare did not begin with the Reagan Administration. In the summer of 1979 the State Department had prepared a detailed compilation of interviews of Laotian refugees on the subject of use of chemical warfare agents. In the fall of that year a U.S. Army medical team visited Thailand to conduct further investigations, and by the winter of 1979 the U.S. Government was raising the matter with the governments of Laos, Vietnam and the Soviet Union. It was also under the Carter Administration that the issue was made public and political pressures mobilized.

108. *Department of State Bulletin*, June 1981, p. 40.
109. *Washington Post*, May 28, 1981, p. 1.
110. William Shawcross, "In Vietnam Now," *New York Review of Books*, September 24, 1981, no. 14, vol. xxviii, p. 62.
111. *Washington Post*, May 1981, p. 1.
112. *Washington Post*, July 14, 1981, pp. A1, 7.
113. Agence France Press Report, October 22, 1981.
114. *Washington Post*, August 11, 1981, p. 1.
115. Prepared statement by Richard Smyster, in *Foreign Assistance Hearings, 1982*, p. 287.
116. *Washington Post*, May 28, 1981, p. 1.
117. Nayan Chanda, "The road to Singapore," *FEER*, September 11, 1981, pp. 9–10.
118. *New York Times*, May 3, 1981, p. 1, 21. *Washington Post*, May 5, 1981, p. 1.
119. *Washington Post*, May 6, 1981, p. 29.
120. Nayan Chanda, "Flip-flop at the villa," *FEER*, August 14, 1981, pp. 20–23.
121. Paul Quinn-Judge, "An arranged marriage," *FEER*, September 11, 1981, pp. 8–9.
122. Nayan Chanda, "Fancy meeting you here," *FEER*, July 24, 1981, pp. 14–15.
123. Nayan Chanda, "Ganging up with exiles," *FEER*, July 31, 1981, pp. 11–12.
124. See prepared statement of Robert P. DeVecchi, Indochina Program Director, International Rescue Committee, in *Foreign Assistance Hearings 1982*, p. 301.
125. *Washington Post*, June 19, 1981, p. 1, 23.
126. "Echoes of Vietnam," *Washington Post*, June 25, 1981, p. A22.
127. Press conference in Manila, June 20, 1981, in *Department of State Bulletin*, August 1981, p. 44.
128. National Public Radio news broadcast, August 18, 1981.
129. *Washington Post*, June 16, 1981, p. A3.
130. Nayan Chanda, "Fancy meeting you here," *FEER* July 24, 1981, pp. 14–15.
131. John Pilger, "American's second war in Indochina," *New Statesman*, August 1, 1980, Vol. 100, No. 2576, pp. 10–15.
132. Michael Richardson, "A really disturbing note," *FEER*, April 10, 1981, p. 23.
133. Nayan Chanda, "A U.N. dove flies into turbulence," *FEER*, March 27, 1981, p. 10.
134. Nayan Chanda, "Sihanouk calls for arms," *FEER*, July 24, 1981, p. 14. Also *Washington Post*, June 18, 1981, p. A36.
135. Nayan Chanda, "Agreement to disagree," *FEER*, July 24, 1981, p. 14. Also, *Washington Post*, June 18, 1981, p. A36.
136. *Washington Post*, July 16, 1981, p. 1.
137. *Ibid*. Also Nayan Chanda, "Agreement to disagree," and Shawcross, "In Vietnam Now."
138. *Department of State Bulletin*, August 1981, p. 37.
139. Holdridge, *U.S. Interests in Southeast Asia*, p. 2.
140. See the statement by U.S. Ambasssador to Thailand Morton Abramowitz in *1980—The Tragedy in Indochina Continues: War, Refugees, and Famine*, Hearings before the Subcommittee on Asian and Pacific Affairs of the Committee on Foreign Affairs of the House of Representatives, 96 Congress, 2nd Session, February 11, May 1, 6, July 29, 1980, p. 124.
141. Haig, *U.S. Interest in Southeast Asia*, p. 3.
142. See, Joseph J. Zasloff and McAlister Brown, "The Passion of Kampuchea," *Problems of Communism*, Jan.–Feb. 1979, pp. 28–43.
143. Nayan Chanda, "Agreement to disagree," *FEER*, July 24, 1981, pp. 13–15.
144. Bernard K. Gordon, *Toward Disengagement in Asia*, Prentice-Hall, 1969.

6

In Search of Peace in the Middle East

Winberg Chai

The momentous crosscurrents sweeping the Middle East could bring radical change throughout the world, as every area of the globe feels its influence. The term Middle East comprises a group of states that are notable for their connections and interactions with one another. J. D. B. Miller of Australian National University has divided the Middle East into five distinct but related subsystems of world politics:[1] First, what one might call the Arab–Israeli group, comprising Israel and its neighbors and traditional antagonists, Jordan, Syria, Iraq, Saudi Arabia, and Egypt; second, the Persian Gulf states, centered on Iran; third, the Horn of Africa and neighboring areas; fourth, those centered on Afghanistan, which include Iran to the south and west, and Pakistan to the east; fifth, the Maghreb states of North Africa, if one stretches it as far as to include Algeria.

One can see how these states relate to one another, often through particular states, as with Iraq and Saudi Arabia between the first and second groups, Iran between the second and fourth, and Egypt between the first and fifth. However, each subsystem has its own problem of interrelationship that is not shared to the same degree by the others. The present chapter is limited to the discussion of the first (Arab–Israeli) group.

THE LEGACY OF THE ARAB–ISRAELI CONFLICTS

The first major Arab–Israeli conflict stemmed from the Arab refusal to accept a United Nations plan adopted on November 29, 1947 to partition Palestine into separate Arab and Jewish states.[2] The United Nations was then a relatively compact body of some 55 members within which the

United States and the Soviet Union were both committed to support a Jewish state. Great Britain, which held a League of Nations mandate over Palestine, ended its mandate on May 14, 1948; and at midnight, the Zionists proclaimed the establishment of the State of Israel. One day later, the armies of five neighboring Arab states—Egypt, Transjordan, Iraq, Syria, and Lebanon—invaded Palestine.

In the ensuing conflict the Israelis successfully defended their new state's existence and occupied 12 of the Arab quarters in modern Jerusalem. They gained about 2500 square miles, approximately 30% more territory than had been assigned to the Jewish state under the U.N. partition plan. Egypt took the Gaza strip, about 135 square miles, while the Jordanian army occupied the Old City and the West bank of the Jordan River.

The Israelis had gained sufficient time and territory to consolidate their state, settling thousands of Jews from the European camps and the Middle East. For the Palestinians, the 1948 war inaugurated the frustration of exile and the despair of stifled ambition. They became the dispossessed of the Middle East.

A second major Arab–Israeli conflict was precipitated by Egyptian President Gamel Abdel Nasser's nationalization of the Suez Canal on July 26, 1956, which grew out of the U.S. decision to withdraw its offer of financial support for the Aswan High Dam. Britain and France enlisted Israel's participation in invading the Sinai Peninsula and drove out the Egyptian troops from the Gaza strip and the Sinai. Universal condemnation of the Anglo-French-Israeli action compelled the withdrawal of the invading forces and the establishment of a United Nations Emergency Force (UNEF) along the Gaza frontier. But Israel gained freedom of navigation through the Gulf of Aqaba. President Nasser attained the pinnacle of his popularity when Syria joined him in 1958 in forming the United Arab Republic.

But the concept of Arab unity proved elusive, and by 1961, Egypt's union with Syria had dissolved. President Nasser wanted to retain a decisive voice in regional affairs and convened the Cairo Conference of 1964, which resulted in the formation of the Palestine Liberation Organization (PLO). From 1962 on, Nasser was enmeshed in an expensive campaign in Yemen against royalist tribesmen who were supported by Saudi Arabia. By 1967, he no longer seemed to be the master of events, and the initiatives in taking action against Israel had fallen to Syria and sections of the PLO. With warnings of an impending Israeli invasion into Syria in May, 1967, President Nasser blockaded the Straits of Tiran to Israeli ships and requested the UNEF to pull out of Egypt.

On June 5, 1967, Israel struck in a lightning move, first destroying the bulk of the Egyptian Air Force, pushing through the Sinai Peninsula to break the Egyptian blockade of the Gulf of Aqaba, and once again put its soldiers on the banks of the Suez Canal. In the east, the Israelis ousted Jordanian troops from the Old City and seized control of all Jordanian territory west of the Jordan River. In six days (June 5–10), Israel established itself as a military power in the region; and unlike in the 1956 conflict, this time its forces refused to withdraw from occupied territories.

Meanwhile, there was worldwide recognition that the region's tensions and conflicts should move toward a solution acceptable to the world community. On November 22, 1967, the United Nations Security Council unanimously approved a resolution, known as Security Council Resolution 242, which called for the withdrawal of Israeli forces from occupied Arab areas; an end to the state of belligerency; acknowledgement of and respect for the sovereignty, territorial integrity, and political independence of every nation in the area; the establishment of "secure and recognized boundaries"; a guarantee of freedom of navigation through international waterways in the area; and a just settlement of the refugee problem.[3]

Most Israelis were relieved that the new territories they had gained from the Six-day War gave the appearance of greater military security. However, military occupation of the West Bank involved Israel with a large Palestinian population at a time when the PLO was evolving from an ineffective body into an aggressive and increasingly sophisticated defender of Palestinian interests. In just 3 years, new fighting was renewed in 1969 along the Suez Canal front. Known as a "War of Attrition," the new offensive was designed to wear the Israelis down and bring about their territorial withdrawals. A year later, another ceasefire was arranged under a new peace plan proposed by U.S. Secretary of State William P. Rogers with a "no-war no-peace" stalemate.[4]

With unprecedented Arab solidarity and newly acquired sophisticated Soviet weapons, the Egyptians and Syrians crossed the Suez Canal and attacked a hard-pressed but resourceful Israeli army on October 6, 1973, the Jewish holy day of Yom Kippur. They broke through Israel's fortifications and advanced into the Sinai Peninsula and the Golan Heights. For the first time, the Arabs used oil as a political weapon against Israel as well as the United States by announcing a 5% reduction in the flow of oil to the United States and other countries supporting Israel. Saudi Arabia announced a 10% cut in oil production and pledged to cut off all U.S. oil shipments if U.S. support of Israel continued. The world's industrial nations were suddenly plunged into an unexpected energy and economic crisis.

Secretary of State Kissinger's "shuttle diplomacy" was instrumental in bringing about a ceasefire between Egypt and Israel on Novemeber 11, 1973. A month later on December 12, the Geneva Conference to discuss Arab–Israel peace was convened in accordance with U.N. Security Council

Resolution 338, which called for the implementation of Security Council Resolution 242. Finally, a second Sinai disengagement pact was signed by Israeli and Egyptian representatives on September 1, 1975.

The lesson of the 1973 crisis revealed that oil had transformed the Arab world into a power of international importance: Saudi Arabia and other oil exporting countries of the Middle East suddenly achieved powerful economic and diplomatic influence.

While at various stages most Middle Eastern countries, as well as the Soviet Union, became involved in intensive negotiations, it was the United States which had become the key outside participant in the Middle East entanglement. By 1977, President Carter became the first U.S. President to endorse the concept of a Palestinian homeland.[5] And the most significant results proved to be bilateral agreements between Egypt and Israel with the U.S. acting as an agent between them at negotiations held at Camp David in September, 1978.

Two important accords were agreed upon: "A Framework for Peace in the Middle East" and "A Framework for an Egyptian-Israeli Treaty," which was finally signed in Washington on March 26, 1979.[6] The Treaty provided for the normalization of relations between Egypt and Israel, which granted Israel her ambition of a bilateral peace with her most powerful antagonist, at the cost of surrendering all her military and civilian gains in the Sinai. The surrendered Sinai oil field when fully developed would aid Egypt's future economic prosperity. Sensitive to the charge that Egypt had asserted her own nationalism to the detriment of wider Middle East issues, the "Framework for Peace" pledged future negotiations on West Bank and Gaza as well as on Palestinian self-rule.

THE EVOLUTION OF REAGAN'S MIDDLE EAST POLICY

During the 1980 Presidential election campaign, candidates from both the Republican and Democratic Parties pledged their traditional strong support for the security of Israel in foreign policy pronouncements pertaining to the Middle East. While the Democratic Platform would have the U.S. continue to play a "full and constructive" role in seeking peace in the Middle East, recognizing its "profound moral obligation" toward Israel, the Republicans, on the other hand, rejected PLO involvement in Middle East peace negotiations.[7]

After assuming the Presidency, Ronald Reagan showed little interest in reviving the Camp David Peace Process originated by President Carter. President Reagan, while personally defending Israel, has left Middle East diplomacy to his staff during the early months in office. To his administration, securing the flow of oil from the Middle East was more important than the Arab–Israeli disputes, the Palestinian problem, or the return of the occupied Arab territories. The Reagan administration wanted to forge an anti-Soviet strategic consensus and partnership between Israel and the moderate Arab states while relegating the bitter regional tension to the background. It was only when the Lebanon–Syrian war broke out on April 2, 1981 and when Syrian SAM missiles were installed on Lebanese soil, threatening a direct Syrian–Israeli confrontation, that President Reagan became fully aware of the volatile Middle East situation.

The Soviet Union, meanwhile, had put its weight behind Syria, Libya, Iraq, South Yemen, and other regional groups. The Soviet invasion of Afghanistan put Soviet fighter jets within striking distance of the Strait of Hormuz, the key to the Arabian–Persian Gulf, through which oil tankers must pass heading for Western Europe and Japan.

On September 17, 1981, Secretary of State Haig spelled out the Reagan strategy in a statement before the Senate Foreign Relations Committee:[8]

> Our broad strategic view of the Middle East recognizes the intimate connections between that region and adjacent areas: Afghanistan and South Asia, northern Africa and the Horn, and the Mediterranean and the Indian Ocean. We recognize that instability in Iran or other areas in the region can influence the prospects for peace between Israel and its neighbors. . . .
> Our proposals to enhance the security of Saudi Arabia are a key element in our Middle East policy. The proposed arms sales will increase the Saudis' ability to defend themselves against local threats; they will directly assist U.S. forces deployed in the region . . . . and they demonstrate our commitment to assist the Saudis against even greater dangers.

Henceforth a definite policy has emerged from the Reagan Administration to arm the Saudis with some of our most sophisticated weapons in order to achieve the "strategic consensus"; and the total list included the following:[9]

1. Five E-3A AWACS (Airborne Warning and Control System) aircraft, at a total cost of $5.8 billion, to be deployed in 1985. Four U.S. Air Force AWACS that have been operating in Saudi Arabia since October 1980 are to remain there until the new AWACS are delivered.

2. 101 sets of fuel tanks that could be fitted to the Saudi F-15s to boost their fuel capacity by 9,750 pounds. Delivery of the tanks, valued at $110 million, was to begin in 1983.

3. Six to eight KC-707 tanker aircraft, worth up to $2.4 billion. The tankers would give Saudi Arabia the capability to refuel both its F-15s and F-5 aircraft

in flight. Saudi Arabia had requested six KC-707s, with an option to buy two more.

4. 1,177 AIM-9L heat-seeking air-to-air missiles, at a cost of $200 million, to replace AIM-9P missiles the Saudis already possessed. The AIM-9Ls would allow the Saudi F-15s to score hits on enemy aircraft flying head-on rather than having to maneuver behind them, as the AIM-9P required.

Although the Reagan administration's strategy also involves closer military relations with Pakistan, Turkey, Morocco, Egypt, Sudan, Somalia, Jordan and Oman, the core of this strategy is Saudi Arabia. The President has assiduously courted the Kingdom as a bulwark for Western defense of the Persian–Arabian Gulf oil fields.[10]

This military strategy is not new. President Carter, concerned with the Iran–Iraq war and the radicalization of Iranian leadership, had already increased U.S. military aid to the Kingdom, including the dispatch of AWACS planes in 1979 and 1980. In addition, the Carter Administration, with the approval of Congress, had sold some 6,500 bombs and missiles to the Saudis for their F-15 warplanes. The U.S. had, in fact, been selling Saudi Arabia military equipment and supplies at the rate of $4 billion to $6 billion a year since 1977. Then in 1978 President Carter persuaded Congress to allow the U.S. to sell the Saudis the 60 F-15s as part of the arms deal that included Israel and Egypt.[11] The total U.S. military sales to the Kingdom in 1969–1980 had exceeded $30 billion.[12]

As early as December 1979, Carter's Under Secretary of State for Security Assistance, Lucy Wilson Benson, had testified before the House Foreign Relations Committee that Saudi Arabia placed "great reliance on its security relationship with the U.S." The Kingdom was worried about "an aggressive Soviet policy in the Middle East. To carry out the policy, the Soviets had aimed at South Yemen, Ethiopia, Iraq, and Afghanistan. The Soviets were zeroing in on these unprotected resources including Saudi oilfields."[13]

There should be little disagreement that a friendly and cooperative Saudi Arabia is crucial to the United States. The economic importance of the Kingdom for Washington even exceeds the strategic gains. It is obviously in the U.S. interest that Saudi Arabia be able to deter other states from launching any type of attacks against it, especially because of the special relationship the United States share with the Saudis, as outlined in a State Department policy paper in September 1981:[14]

> Saudi Arabia traditionally has been the most moderate OPEC state and consistently has shown concern for the world economy. It has maintained oil production higher than preferred levels – increasing production to 10 million barrels per day to offset shortages resulting from the Iran–Iraq war – and priced Saudi crude well below the general OPEC level. Saudi Arabia has devoted nearly 10% of its GNP to foreign assistance to such

moderate Arab and Islamic states as Morocco, Turkey, Pakistan, Sudan, Oman, Jordan, and Bahrain, in many cases complementing U.S. efforts.

The Saudis have been a principal force in countering Soviet efforts to increase their influence in the region. They led the Arab world in condemning the Soviet invasion of Afghanistan and have assured that this subject receives priority treatment in Arab and Islamic Councils. Saudi Arabia also has played a constructive role in securing the current cease-fire in Lebanon, complementing U.S. diplomacy.

Moreover, the sale of AWACS by the Reagan Administration was a continuation of the U.S. defense strategy in the region during the past several years. In addition to military weapon sales, there are less known but equally important elements of the U.S. involvement with the Kingdom: the activities of the U.S. Army Corps of Engineers, totalling $24 billion, including construction of a self-contained military base and city in the middle of the desert; the Saudi naval expansion program, which is a $6 billion effort to build the Saudi Royal Navy; and the $4 billion Saudi Arabian national guard program, etc. The ultimate goal has always been the establishment of a "regional wide air-defense network", led by Saudi Arabia and potentially including such moderate states as Kuwait, the United Arab Emirates, Oman, Bahrain, and Qatar.[15] The Saudis have already taken the lead in the formation of the Gulf Cooperation Council (GCC) with a major objective to enhance the defense of the Gulf.

At the moment, Reagan's Defense Department is working on a regional defense strategy to link the air defense networks of the Gulf states with the United States into a unified system, a continuation of the Carter defense planning in 1980.[16]

First, Mitre Corp., a research firm that does much of the U.S. Air Force advanced electronic work, completed a 2-year study for the Saudis in 1980, suggesting how the latest in computer technology could be programmed to coordinate Saudi air defenses and if needed, all U.S. forces in the area.

Two other studies examined plans to pull all three under one integrated command, control, and communication, known as the C^3 system. The U.S. can easily hook into a Saudi C^3 system, even when not operating directly with the Saudis, and can get all the same data simultaneously transferred by satellite both to U.S. bases in the Indian Ocean and to the Pentagon.

Moreover, the advanced C^3 system can also integrate electronic intelligence information, creating a command, control, communications, and intelligence system, known as a C^3i (or cubed eye) system. With this new technology, battlefield commanders will be able to calculate almost instantly the precise location of enemy ground, air, and naval forces, and to target them with the most efficient and effective battle plan from computer display consoles in a central command post. It is President Reagan's hope that by 1990 this C^3i system will link other components of the program into an integrated combat network for the defense of the Saudi Kingdom and other Gulf states.[17]

President Carter in 1980 did not announce nor push this strategic arrangement with Saudi Arabia for fear that the U.S. supporters of Israel would oppose it, especially in an election year. According to one Carter official, "Israeli Prime Minister Begin sees the long-term trend toward U.S. security relations with Arab states as being at the expense of Israel's political security and eventually, maybe, military security... Begin wants to force the Arabs to come to terms with Israel on Israel's terms."[18]

In 1981 we witnessed the most strenuous battle in the U.S. Congress in recent history which pitted the nation's Jewish community and its powerful pro-Israel lobbying organizations and sympathizers against the executive branch with regard to the AWACS sale to Saudi Arabia. President Reagan conceded the House even before the formal debate began; he personally lobbied in the Senate. The Israeli bombing in Iraq and Lebanon in the summer months, and the assassination of Egyptian President Sadat on October 6, shifted the focus of the AWACS debate and gave President Reagan new hopes of winning in the Senate. On October 28, before a standing-room-only gallery, the Senate voted 52–48 in a narrow margin, granting a major foreign policy victory for President Reagan.[19]

AWACS sale victory in Congress would have been a dramatic turning point in U.S. relations with the Arabs had President Reagan seized this opportunity to push Israel to the negotiating table based upon a dramatic new proposal initiated by Crown Prince Fahd of Saudi Arabia in August 1981. The essentials of the Saudi Peace Plan included the following:[20]

1. Israeli evacuation of all Arab territories seized during the 1967 Middle East war, including the Arab sector of Jerusalem.

2. Dismantling the settlements set up by Israel on the occupied lands after the 1967 war.

3. Guaranteeing freedom of religious practices for all religions in the Jerusalem holy shrines.

4. Asserting the rights of the Palestinian people and compensating those Palestinians who do not wish to return to their homeland.

5. Commencing a transitional period in the West Bank of Jordan and the Gaza Strip under United Nations supervision for a duration not exceeding a few months.

6. Setting up a Palestinian state with East Jerusalem as its capital.

7. Affirming the right of all countries of the region to live in peace.

8. Guaranteeing the implementation of these principles by the United Nations or some of its member states.

The Crown Prince's Peace Plan stressed Arab political and nationalistic terminology instead of emotional reactions. The Plan was put in such a form as to offer the U.S. and Western Europe a greater degree of flexibility, while it denied the Soviet Union any direct involvement. It, in fact, did not mention the Palestine Liberation Organization by name, but rather referred to the Palestinian people as a whole. This is an important concession since Israel has refused to negotiate directly or indirectly with PLO. Most important of all, the Plan implies for the first time, by any Arab state, the recognition of the state of Israel.

Western European reaction to the proposal was generally favorable. Lord Carrington, Britain's Foreign Secretary, called it "a departure and very encouraging."[21] French Foreign Minister Claude Cheysson attached "great importance" to the concept.[22] West German leaders referred to the new proposal as "interesting."[23]

Predictably, Prime Minister Begin of Israel lost no time in expressing his opposition. He wrote a personal letter to President Reagan denouncing the plan, and publicly called the Fahd Plan "a plan designed for Israel's liquidation."[24] Following Lord Carrington's endorsement of the proposal, Israeli Foreign Minister Yitzhak Shamir vowed to exclude from the planned Sinai peace-keeping force "any European government that undermined the Camp David Peace Process."[25] To make matters worse the Israeli cabinet, in an unprecedented move, annexed the Golan Heights on December 21, 1981.

As indicated earlier in this chapter, the Reagan Administration seemed to take little notice of the Saudi proposal when it was first announced. But after the U.S. Senate approved the AWACS sale, some policy makers in the State Department let it be known that the Crown Prince's peace plan was under "active discussion."[26] Dean Fischer, the U.S. State Department Spokesman, said that Prince Fahd's statement implicitly recognizes Israel's right to exist and that Washington welcomes "any contribution to achievement of a just and comprehensive peace in the region."[27]

The demise of the Saudi peace proposal was caused as much by the lukewarm reception of its ally, the United States, as it was vetoed by radical Arab states in the Arab League. Yasir Arafat of the PLO welcomed the proposal, but the radical groups of Georges Habash and Naif Hawatme, supported by Syria, rejected it.[28] President Anwar el-Sadat of Egypt was also not in favor of the Plan because it took no note of the Camp David Accords. Although both King Hassan of Morocco and King Hussein of Jordan expressed their support, they were unable to convene an Arab League Summit at Fez in 1981 to endorse it. After the collapse of the Fez Conference, the Reagan Administration was known paradoxically to be somewhat relieved.

THE REAGAN PEACE PLAN

In the violent and bewildering kaleidoscope of the Middle East, moments occasionally appear when the political pieces are so shaken that old patterns are irrevocably altered, paving the way for new and unpredictable trends. The Lebanon invasion by Israel in the summer of 1982 may be one of these events.

First of all, whatever gains were made since 1975 by the PLO under the tutelage of the Soviet Union, and by Communists, socialists, and Nasserites in Lebanon, have now been firmly checked by the Israeli incursion. In the Winter of 1982, the U.S. moved to center stage and opened up extraordinary opportunities for a dynamic American diplomacy throughout the Middle East. As Henry Kissinger commented during the Lebanon War:

> Events in Lebanon should enable us to overcome the existing fragmentation of our policy and to relate in a comprehensive approach the three great issues of the Middle East: the Lebanese crisis; the autonomy talks regarding the West Bank and Gaza, and the threat to Western interest in the gulf.[29]

Second, the failure of the Soviet Union to evince support for the beleaguered PLO and Syrian forces during the early stages of the Israeli invasion reflects "mutual Soviet–Arab disenchantment."[30] According to Ambassador Talcott W. Seelye, Soviet disenchantment with Arabs has been based on the following factors:[31]

1. Sadat's precipitate expulsion of Soviet advisers in the early 1970s after years of close relationship.
2. Syria's entry into Lebanon in 1976 over strong Soviet objections.
3. Iraq's ignoring of Soviet objections to launching a war against Iran and its severe treatment of Iraqi Communists.
4. Algeria's turning away, under President Chazli Ben Chadid, from a previous close Soviet embrace.
5. The continued refusal of Saudi Arabia, Oman, Behrain, the U.A.E. and Qatar to permit a diplomatic Soviet presence.
6. Finally, the mercurial and unpredictable behavior of one of the Soviet Union's supposedly best clients, Libya's Muammar Qaddafi, with whom the Soviet Union cannot be fully at ease.

From the Arab perspective, the Soviet brand of socialism lost its appeal as heavy injection of it had failed to remedy the ills of Arab society. In fact, again according to Ambassador Seelye, Arab economics became worse off than they had been under the much maligned entrepreneurialism. Moreover, most Arab governments, with perhaps one exception being South Yemen, had welcomed Soviet assistance and close cooperation, but fiercely resisted Soviet efforts to interfere. Often the Soviets acted heavy-handedly and occasionally bumbled. Bitter Arab experience with the Soviet Union over the years dispelled earlier illusions about them.

As indicated before, when he was elected, President Reagan brought no magic formula for the problems of the Arab–Israeli conflict. The Lebanon invasion, with the introduction of U.S. marines in Beirut, moved the President from the role of a passive "mediator" to that of a "full participant" in the search for peace in the Middle East. He saw new opportunities for the United States as he called for a "fresh start" in the region in a televised speech on September 1, 1982, the day that the PLO withdrawal from West Beirut was completed.

The President reaffirms "ironclad" U.S. support for Israel and the Camp David peace process. He expands the context of the Camp David accords beyond the narrow definition favored by Israel; and he asks that the 5-year process prescribed in the 1978 accord to bring autonomy to the Palestinians on the West Bank be commenced. That process has never been initiated because of Prime Minister Begin's misgivings about it and continued resistance by the PLO as well as by Jordan.

President Reagan sees the Arab–Israeli conflict as a problem of reconciling Israel's legitimate security concerns with the legitimate rights of the Palestinians. He apparently wishes to introduce new ideas strongly resisted by Prime Minister Begin in the past. In summary, President Reagan has initiated a positive U.S. policy calling for the acceptance by all parties concerned of the following "principles"[32]:

1. Full autonomy over their own affairs for Palestinian inhabitants of the West Bank and Gaza.
2. Five-year period of transition as outlined in the Camp David process.
3. Self-government by the Palestinians of the West Bank and Gaza in association with Jordan.
4. A settlement freeze by Israel that would preclude further Jewish settlement in the occupied area.
5. Opposition to the formation of an independent Palestinian State in the Israeli-occupied West Bank and Gaza Strip.
6. Opposition to annexation or permanent control by Israel in these areas.
7. Jerusalem must remain undivided and its final status should be decided through negotiations.
8. Acceptance of U.N. Resolution 242.

It can be said that President Reagan is being swayed by the Lebanese events back into the mainstream of U.S. foreign policy as worked out by his three predecessors, Richard Nixon, Gerald Ford, and Jimmy Carter. After some 18 months in the White House, President Reagan realizes a comprehensive peace settlement in the Middle East is an urgent and necessary task.

This is a monumental challenge for any U.S. President, but failure could bring unthinkable calamities to the industrial world of the West and Japan. It could insure the radicalization of the remaining moderate Arab states and of the Palestinian populations. Since their crushing defeat by Israel in 1967, most Arab governments have been fighting a rearguard action to cling to power. The resurgence of Islamic fundamentalism in this period is "in large part a response to the failure of these regimes to fulfill their promise for national strength."[33] Leaders of these moderate Arab governments are fearful that unless there is a comprehensive peace settlement, there may be a violent backlash throughout the Arab world that could threaten many governments, both moderate and radical.

In the past, the U.S. has always been able to count on Saudi Arabia to help bring peace and stability, as they demonstrated in the case of the difficult negotiations in Beirut. The Saudis have played a major behind-the-scenes role in nudging the Palestinians toward a formula for their exit from Beirut that would also be acceptable to the Israelis.[34]

In the 1982 Arab League summit conference at Fez, Morocco in September, the Saudis again scored a major victory by pushing through the adoption of King Fahd's Peace Plan as a joint declaration of the Arab League. A seven-member delegation headed by King Hassan of Morocco representing the Arab League travelled to the U.S. and presented the Plan to President Reagan in October 1982. In it there is a key section which calls on the United Nations Security Council to guarantee "peace among all states of the region including the independent Palestinian state."[35] This certainly suggests that the Arab states are ready as a group to recognize and coexist with Israel.

President Reagan's new peace offensive has touched off a deep debate in Israel. While the government of Menachem Begin has arrogantly rebuffed his initiative, leader of the Labor Party, Shimon Peres, for example, has enthusiastically endorsed the American proposal as "a most realistic basis for negotiations and for the continuation of the peace process in the Middle East, and therefore it is a great asset."[36]

Rarely has an American President had more potential backing and leeway for putting pressure on Israel. Mr. Reagan has subtly placed the entire American–Israeli relationship in the balance by warning that "if Mr. Begin continues to act unilaterally, he will find himself alone, without American support."[37]

"War changes conditions, but it does not solve problems or make peace," wrote a retired U.S. diplomat.[38] The 34-year-old Arab–Israeli conflicts are heavily loaded with much emotion, prejudice, ideology, apprehension, and bloodshed which make their solution extremely intricate. To convince Israelis and Palestinians that honest negotiations are possible, each side must be persuaded that the other is ready to negotiate.

President Reagan, in his September 1982 address to the nation, made three urgent appeals:[39]

> I call on the Palestinian people to recognize that their own political aspirations are inextricably bound to recognition of Israel's right to a secure future.
> I call on the Arab states to accept the reality of Israel and the reality that peace and justice are to be gained only through hard, fair, direct negotiation.

But President Reagan's most important call was to the state of Israel, the victor in the Lebanon invasion: "I call on Israel to make clear that the security for which she yearns can only be achieved through genuine peace, a peace requiring magnanimity, vision, and courage."[40]

The Reagan initiative, to be sure, has more than a few shortcomings in Arab eyes, as it calls for the creation of a Palestinian entity in the West Bank and Gaza strip in association with Jordan, falling short of an independent Palestinian state. Whatever its deficiencies, however, many Arabs are beginning to view the Reagan plan as the first U.S. peace initiative that breaks out of the Camp David framework and meets their minimum demands for the return of Israeli-occupied territories. As such, it is seen by the Arabs and Palestinians as an advance on Camp David and is being welcomed accordingly.[41] The Fez Declaration is their olive branch in response to the Reagan peace call.

Nevertheless, nothing is likely to be done unless President Reagan, like President Eisenhower during the Suez crisis, insists that Israel change its policy, at pain of losing all U.S. economic and military aid. Yet the sad thing is: As long as Begin remains at the helm of the Israeli statecraft, fighting on his crutches to the end, *mere* threats from Reagan are unlikely to produce the necessary change, even in the face of strong demands by the press and people of Israel that Begin reappraise his policy following the Lebanese invasion and massacre.[42]

Mr. Reagan needs to take additional steps to protect U.S. interests in his dealings with Israel, for example, the postponement of 1983 U.S. aid, about $2.2 billion, in order to induce Mr. Begin to compromise. In early October 1982, with the final passage of the fiscal year 1983 Continuing Appropriation Resolution, Israel held the all-time record with a cumulative total of $25.3 billion in U.S. aid, about one-tenth of all aid dispersed by the United States abroad since Israel's creation in 1948.[43]

The President also needs to assure Begin that the United States will not sacrifice Israel's security needs in the years ahead. President Reagan can insist on a permanent armament freeze in the region including the neutralization of the West Bank and Gaza Strip in association with Jordan as a future homeland for the Palestinians.

NOTES

1. J. D. B. Miller, "A Region of Constant Surprise," In M. Ayoob (ed.) *The Middle East in World Politics* (New York, St. Martin's Press, 1981), p. 203.
2. The brief historical background presented in this section is based on the documentary history by T. G. Fraser (ed.) *The Middle East, 1914–1979* (New York: St. Martin's Press, 1980). The two best documentary sources, other than Fraser's work, are *The Middle East* (Washington, D.C., Congressional Quarterly, Inc.) 5th ed., 1981 and 4th ed., 1979.
3. The Palestine refugee problem is the subject of many books and articles, most of which are political and emotion-charged. However, the best two accounts to appear recently are William B. Quandt, Fuad Jabber, and Ann Mosely Lesch, *The Politics of Palestinian Nationalism* (Berkeley: University of California Press, 1973); and Ibrahim Abu-Lughod (ed.) *The Transformation of Palestine* (Evanston, Ill.: Northwestern University Press, 1971).
4. *Department of State Bulletin*, Vol. 67, No. 1593, January 4, 1970.
5. See chapter on "No More War," in Jimmy Carter, *Keeping Faith* (New York: Bantam Books, 1982), pp. 267–429.
6. For a pro-Israel position, see Michael Curtis, "Introduction after the withdrawal from Sinai," *Middle East Review*, Vol. XIV, Nos. 3–4, Spring–Summer 1982, pp. 3–4.
7. *Congressional Quarterly Weekly Report*, August 16, 1980, p. 2365.
8. *Department of State Bulletin*, Vol. 80, No. 2055, October 1981, p. 14.
9. *The Middle East*, 5th ed., p. 61.
10. *New York Times*, November 4, 1981.
11. *The Middle East*, 5th ed., p. 59.
12. *Ibid.*, p. 49.
13. *Ibid.*, p. 59.
14. "Air Defense Equipment for Saudi Arabia," *GIST*, September 1981.
15. *Washington Post*, November 1, 1981.
16. *Ibid.*
17. *Ibid.*
18. *Ibid.*
19. The intensity of the Jewish lobby against the AWACS sale can be illustrated from an article published by the *Chicago Tribune* of an interview by U.S. Congressman Dan Rosetenkowski, who told reporters that he voted against selling the AWACS, although he personally "believes the sale is right", however, he did not want "Jewish groups coming down on me." *Chicago Tribune*, October 17, 1981; also see *Congressional Quarterly Weekly Report*, Vol. 39, No. 44, October 31, 1981.
20. *New York Times*, October 31, 1981.
21. *Newsweek*, November 16, 1981, p. 45.
22. *MacLean's*, September 21, 1981, p. 3.
23. *Mideast*, August 19, 1981, p. 43.

24. *Time*, November 16, 1981, p. 24.
25. *Newsweek*, November 6, 1981, p. 44.
26. *Ibid.*, p. 41.
27. *New York Times*, October 31, 1981.
28. *Al-Hadaf*, August 25, 1981.
29. *Washington Post*, June 16, 1982.
30. *Christian Science Monitor*, August 11, 1982, p. 22.
31. *Ibid.*
32. *Department of State Bulletin*, Vol. 82, No. 2061, September 1982, pp. 23–25.
33. John Waterbury, "Arabs on Edge", *New York Times*, November 9, 1982.
34. *Christian Science Monitor*, August 12, 1982.
35. *New York Times*, October 24, 1982.
36. *Washington Post*, September 12, 1982.
37. *New York Times*, September 26, 1982.
38. Harold H. Saunders, "Post Lebanon Goals," *New York Times*, June 20, 1982; also see his article on "An Israeli–Palestinian Peace," *Foreign Affairs*, Vol. 61, No. 1, Fall, 1982, pp. 100–121.
39. *U.S. Department of State Bulletin*, Vol. 82, No. 2061, September 1982, pp. 23–25.
40. *Ibid.*
41. Thomas L. Friedman, "New Realtities, Old Enmities Shake Up Arab Alignments," *New York Times*, October 17, 1982.
42. James Reston, "The Tragedy of Begin," *New York Times*, September 22, 1982.
43. Alan C. Kellum, "U.S.–Israeli Relations: A Reassessment," *The Link*, Vol. 15, No. 5 (December, 1982), pp. 2–3.

7

Invisible Enemies: Conflict and Transition in Soviet–U.S. Relations

Paul H. Borsuk

As the heralded year 1984 approaches, the two primary powers in world politics appear to have arrived at a point in their relationship where there is in fact almost no relationship at all. At the same time, neither the United States nor the Soviet Union now appears to wield concerted influence in any third country or region, even those routinely deemed vital to its interests.

Were it not for the fact that these two states alone could bring about thermonuclear catastrophe as they practice their respective versions of global involvement, these observations—if valid—might only be absurd. But in light of the possible consequences, serious analysts should perhaps be searching for answers to some rather insolent-sounding yet very urgent questions. How and why did U.S.–Soviet relations disappear? Who was involved in the disappearance and how reliable are the witnesses? Where and under what circumstances are they likely to reappear? What are some of the possible consequences of this nonrelationship for basic patterns of world politics for the balance of the 1980s and beyond?

This chapter does not offer a detailed narrative of the twists already taken by Soviet–American relations over the two decades since the Cuban missile crisis.[1] Nor does it constitute a polemic advocating one defined course of action over another for Washington to take toward Moscow. Instead, in a highly interpretive fashion it will explore the thesis that

---

The views expressed in this chapter are solely those of the author. Reference to his employment by the U.S. Central Intelligence Agency is for personal identification only. This material has been reviewed by the CIA to ensure that no classified information has been included. This review neither constitutes CIA endorsement of the author's views nor implies CIA authentication of the content as factual.

U.S.–Soviet relations may have come to a temporary halt mainly for reasons that relate to each country's internal political development rather than issues such as Afghanistan, Poland, or strategic arms negotiations. This chapter tries to point out some of the questions of basic circumstance in Soviet–American relations that will necessarily find answers as Moscow and Washington reassess their respective positions in world politics and their goals and accomplishments at home.

NOTE ON THEORETICAL CONTEXT

To phrase this thesis a bit differently, this study contends that both the Soviet Union and the United States are now passing through a period of intense conflict and change within their respective dominant political cultures, with special relevance to foreign policy. Political culture, by this reckoning, is not a superficial phenomenon. Rather it constitutes an entire cultural subsystem that defines a group's collective basic expectations and assumptions about politics, including international politics. This group may at times coincide with the entire populace of a single state, or may refer instead to a defined elite or cluster of elites. In some contexts, the terms "political culture" or "elite political culture" can be substituted for the term "ideology"; in other contexts, the term "myth" may do as well. In any case, primary reference is to a consensus that unifies the political actions of large numbers of individuals while they act as members of complex, functionally autonomous organizations, sometimes only loosely subordinated to the state.[2]

The conscious and diligent cultivation of this consensus is one of the primary tasks of leadership in any complex organization, including the state itself.[3] Success at this task provides one of the primary avenues toward long-term institutionalization of such an organization. Institutionalization here refers to much more than an artificial concept created by and for social scientists. In concrete terms, institutionalization is the process that transmits and preserves elite political culture, without dependence on any one living individual. The result is creation of collective attitudes and perceptual tendencies pertaining to large numbers of individuals simultaneously—the very factors that ultimately enable scholars and other outside observers to speak meaningfully about entire entities such as governments or political movements as though they were individual actors.

Dominant political culture within an institution has much to do with development of the learning ability, or adaptive qualities, that the institution will eventually have to exhibit as it conducts relations with competing institutions and other forces in the external environment. A stable political culture, able to combine success in internal consensus building, with subsequent success in external information handling, can reap a particularly precious political harvest—political legitimacy.

Awareness of these linkages can help resolve the analytic task of understanding how an enormous entity like the nuclear-armed state functions at all, and how individuals with no direct knowledge of each other can routinely act in concert as members of a single institution. Captured with the layers of bureaucracy is a central consensus about the purposes of politics. If this consensus is especially elaborate or is explicitly taught as a philosophy of truth, it might be called an ideology; if it is especially assimilated in the institution's routines or able to depend on a few universal symbols and legends, it might better be called a myth. In either instance, such consensus is what stubbornly breathes life into abstractions such as the national security state or the "military–industrial complex."

When this consensus is impaired or dissolved, the consequences for state behavior can be considerable. Fortunate indeed is the country that can pass through such an interval of conflict over political, economic, and cultural fundamentals—a conflict that may decide how national security is to be redefined and how scarce resources are to be reallocated accordingly—without suffering serious losses or becoming involved in some form of violence with another state. Thus far through the early 1980s, this has been the good luck of both the superpowers. By the end of the decade, however, both of the superpowers may be ready to resume direct competition in defined areas of mutual interest—animated by a new, perhaps improved, myth or sense of collective purpose.

In the hope that none of this has tried the reader's patience too much, this essay will now turn to specific questions of how and why Soviet–American relations entered a period of apparent dormancy, and how that period might end.

HOW DID U.S.–SOVIET RELATIONS VANISH?

In retrospect, it seems clear that the United States and the Soviet Union each believed that the agreements negotiated between 1972 and 1974 would eventually reduce the other party's freedom of action, at little or no cost to the first party. For the United States, Moscow's growing dependence on a steady influx of Western goods and services would induce a desire to accomodate Western expectations on a number of noneconomic issues, at the same time that the basic structures of the Soviet system would be compelled despite themselves to change their fundamental character. For the Soviet Union, Washington's need to preserve its post-Vietnam claim that a global design built on negotiation rather than intervention was not just successful but triumphant would insure that squalid little quarrels about which of the competing superpowers would claim Angola as an ally would never intrude on expanding high-level dealings.

But by mid-1975, leaders on both sides had begun to realize that their assumptions had been faulty, and both the orbital linkup of astronauts and cosmonauts and the ceremonial signing of the Helsinki accords on European security and cooperation became the grandest of empty gestures.[4] Yet responses remained for the most part restrained. Washington may have governed its response on the grounds that the Vladivostok agreement on strategic arms parity might take on life of its own, at the same time that Bonn should be permitted to broaden its recent successes in Ostpolitik. Soviet poise in the face of the deliberately provocative Jackson–Vanik amendment, embodied in the Trade Reform Act of 1974, may have been prompted by a combination of fresh unease over China's new internal political stability and desire to preserve the Nixon foreign policy legacy of business-like dealings with all comers.

Public confidence inside the United States in the utility of a working relationship with the Soviets began to erode seriously in the heat of the 1976 Presidential campaign. The uproar that followed President Ford's impromptu assertion that the fruits of detente included an end to Soviet domination of Poland turned a human error into a symbol both of official bungling and of Soviet perfidy, ostensibly unexpected. Not even the ceremonial removal of the term detente from the Administration's "political lexicon" could lend new glamour to the dry path of negotiation with Moscow over the mundane things that divide any two competing states.

President Carter appeared sincere in his beliefs both that major issues between the superpowers could be resolved quickly and decisively and that the agenda of world politics was about to change to reflect new realities of North–South economics rather than East–West politics. In this spirit, he sent Secretary of State Vance to Moscow in March 1977, confident that his abrupt change of goals and tactics in strategic arms negotiations would galvanize rather than alienate the Soviets. The contrary was the case, however, and by 1978 a trilateral variant on the Nixon–Kissinger leverage theories that was much less deferential to Soviet sensitivities had been formulated. To make up for the debacle of the "neutron bomb," NATO would achieve and implement an accord on real increases in military expenditures; similar cooperation would be accelerated with Japan. Most significantly, the normalization of relations with post-Mao China would be pursued with fresh vigor—and eventually given precedence over conclusion of a Vladivostok-based strategic arms pact with the Soviet Union.[5]

To avoid antagonizing Moscow in every arena except the strategic arms talks, the Administration in October 1977 invited the Soviets to share credit for the peace in the Middle East that was sure to come; price of admission would be the Palestine Liberation Organization, surely a bargain in light of the PLO's functional incompetence except in some sectors of global public relations. This initiative was promptly rendered meaningless to Moscow by the direct, ultimately fruitful diplomatic exchanges that followed Egyptian President Sadat's spectacular visit to Jerusalem. Washington, on the other hand, could concentrate on expanding its involvement—which was widely thought to connote influence—in the negotiations.

Meanwhile, a particularly nasty debate had begun inside the United States on the proposition that a massive compound failure of intelligence and diplomacy had taken place in the dealings with Moscow that seemed so universally popular just a few years before. At first limited to a few scholars and analysts, prominently including the members of "Team B" who were officially summoned to reprocess the data on Soviet missile development, the dispute about Soviet intentions soon spilled over into the chemistry of the approaching Presidential election campaign of 1980—to the expanding advantage of the Republican challenger.

Almost unnoticed, the SALT II treaty was signed by the heads of state in Vienna in June 1979.[6] The start of a very perilous ratification process for the document was first delayed by the controversy over the size and capability of the Soviet military presence in Cuba, then postponed indefinitely by the Soviet military intervention in Afghanistan at the end of 1979. Whether such a move on Moscow's part would have been attempted if the outlook for relations with the United States were viewed more positively is impossible to guess, just as the question of whether detente at its height could have withstood the invasion if it had taken place cannot be answered. In the words of Seweryn Bialer, the Soviet Union prior to its Afghan decision apparently considered that it had nothing either to hope for or to fear from Washington. A more concise description of the cessation of U.S.–Soviet relations on the eve of the 1980 Presidential elections would be difficult to find.[7]

The Reagan difference in the U.S. approach to dealings with the Soviet Union is best expressed in the President's own words. Questioned at the outset of his Administration about his views of Soviet intentions over the long term in world affairs, he responded:

I know of no leader of the Soviet Union since the revolution, and including the present leadership, who has not, more than once, repeated in the various Communist congresses [that] they hold their determination that their goal must be the promotion of world revolution and a one-world Socialist or Communist state, whichever word you want to use.[8]

On the same occasion, President Reagan characterized the grain embargo imposed on the Soviet Union following the Soviet invasion of Afghanistan at the end of 1979 as "more of a kind of gesture than it was something real."[9] Sanctions might be levied against Moscow in the future, but these would not require "one group of Americans to bear the burden."[10]

As enunciated by the President during his June 8, 1982 address to a joint session of the British Houses of Parliament, U.S. policy should henceforth be built on an awareness that a profound internal crisis is unfolding in the Soviet Union. The President rejected as "preposterous" the argument that "we should encourage change in right-wing dictatorships but not in Communist regimes."[11] The primary means toward encouraging positive change within totalitarian states, according to the President, consists of an effort "to foster the infrastructure of democracy—the system of a free press, unions, political parties, universities—which allows a people to choose their own way, to develop their own culture, to reconcile their own differences through peaceful means."[12] Contrasting long-established direct cooperation among Western European political institutions to past Soviet efforts to promote political violence and subversion abroad, President Reagan declared that the United States intended to take "concrete actions" to strengthen democratic institutions and practices internationally, in consultation with U.S. allies. Ultimate success in this effort, though it might be long in coming, "will leave Marxism–Leninism on the ash heap of history."[13]

In the same address, the President reiterated the "firm and unshakable" U.S. commitment to an early successful outcome in the Strategic Arms Reduction Talks (START).[14] One month earlier, in his May 9 address to the graduates of Eureka College, President Reagan elaborated on the theme of negotiations with the Soviet Union. With specific regard to START, he said:

> We will negotiate seriously, and in good faith, and carefully consider all proposals made by the Soviet Union. If they approach these negotiations in the same spirit, I'm confident that together we can acheive an agreement of enduring value that reduces the number of nuclear weapons, halts the growth in strategic forces, and opens the way to even more far-reaching steps in the future.[15]

Also in his Eureka address, the President declared:

> ...a Soviet leadership devoted to improving its people's lives, rather than expanding its armed conquests, will find a sympathetic partner in the

West. The West will respond with expanded trade and other forms of cooperation. But all this depends on Soviet actions. . . . [16]

Millions of Americans demonstrated by their votes in the Presidential election of 1980 that their distrust of the Soviet Union as an international negotiating partner had reached new heights. Early in the new Administration, however, millions of Americans—possibly including many whose votes had been for the victor in the 1980 campaign—were expressing in a variety of ways their lack of support for weapons procurement programs and other military initiatives deemed by some to be an essential part of a comprehensive answer to the Soviets. As 1982 ended and concern about the medium-range outlook for the American economy became more widespread, there was little sign that this contradiction in basic public policy would soon be resolved.

In the meantime, the Soviet Union was enjoying some diplomatic satisfaction in the course of opening narrow but steaming rifts between the United States and its West European NATO partners over related issues of missile diplomacy and commercial policy toward Moscow. But by this time it had become clear that no amount of commerce and technology from the West would reverse the trend toward industrial stagnation and agricultural decline that pervaded the Soviet economy. The avoidance of full-scale military intervention in the Polish crisis represented a major nonevent in Soviet policy making, representing an immense saving in terms of both political and economic cost—but offering nothing remotely resembling a long-term answer to the challenge of redefining the Soviet Union's responsibilities and expectations in Central and Eastern Europe.

With more than 100,000 Soviet troops thrashing through Afghanistan trying to win a political war, with detente with the United States a memory but with no defined relationship yet having taken its place, Leonid Brezhnev died. At the same time, the U.S. President who signed Brezhnev's condolence book was beset by new demands from within that he change course both at home and abroad, or risk losing the reelection campaign he might choose to wage. Not since Stalin's death had both the United States and the Soviet Union stood simultaneously at such a clear crossroads.

SOVIET POLICY AND POLITICAL CULTURE: REALITY IN SEARCH OF MYTH

Yuri Andropov's elevation as Leonid Brezhnev's successor may be signal that the Soviet Union is preparing to make long overdue choices among contradictory political commitments, economic strategies, and ideological claims. At the very least, Moscow's policies on a wide variety of foreign and domestic issues seem likely to enter a period of maneuver, if only to dispel the impression abroad (and at home) that Soviet institutions and their leaders have become totally immobile.[17]

Andropov's political character remains to be demonstrated—as does his health in the highest office. But it seems safe enough to call him the best informed man in Russia. Unlike most of his rival claimants to leadership over the Soviet party and state, he has been professionally obliged, day in and day out, to accept and act on adverse information about Soviet performance at home and abroad. If he has done his job at all adequately over 15 years as chief of the KGB, he should have few illusions about the extent of corruption, inefficiency, and incompetence in the Soviet economy, about the failure of the Soviet military to pacify Afghanistan, or about the futility of political exhortations intended to cause the Polish economy and the Polish Communist Party as well to revive and prosper.

In beginning any fresh effort to think about Soviet intentions and capabilities, it may be time to abandon the belief that a single leader—or even a single, table-sized committee—effectively rules the Soviet Union. This is not to suggest that Soviet political power has become shared or diffuse among millions. Instead, the need may now be to monitor the interactions not just of a few dozen individuals but of a few dozen institutions—each one with a life of its own and a consensus of its own, yet each a member of a larger institution, namely the Soviet system itself. In other words, the premise to be studied here is that the Soviet Union is well on its way to becoming a mature, self-stabilizing oligarchy.[18]

This is to suggest that the arena of Soviet politics so often scrutinized by "Kremlinologists" and other scholars from abroad may increasingly consist of a series of arenas, where the rules of competition over resources and privileges emerge and shift in an increasingly orderly way. The same senior forum may process different issues differently, as when the Presidium of the Council of Ministers bitterly debates an annual economic plan for an industrial sector and then formally approves a program of military procurement already agreed to elsewhere. Conversely, the same issue can be discussed in very different ways in nominally identical settings, as when a multitude of managerial entities must discuss the same revised procedure or plan. From another perspective, it may be possible that only 15 men give final approval to grand strategy proposed toward the United States or China, and that 15 men likewise are responsible for assessing the progress of Soviet diplomacy toward some of the small states of the third world. Clearly these will not be the same 15 men, yet they will act with similar self-perceptions of genuine professional responsibility.

In short, a focus on the behavior and interaction of organizations may have as much or more to contribute to an understanding of Soviet politics as a focus on the aims and characteristics attributed to individual political leaders.[19] By the end of the 1980s, regardless of the length of Andropov's tenure as General Secretary and the identity of his successor, the Soviet polity will likely appear an aggregate of professional and political elites responsible for the performance of more than just a few major Soviet institutions. By even a narrow definition, the number of individual personalities involved could number not just dozens but hundreds, even thousands.[20] In such a context, such factors as interelite and intraelite consensus, the evolution or continuity of mutually apparent political culture, the emergence of counterelites that advise rather than dissent — all these will have more to do with the elemental analysis of collective Soviet behavior, in internal politics as well as foreign policy, than earlier observers would have imagined.

The prevailing consensus or shared self-image among these hundreds or thousands of political participants might even take on some of the aspects of an unwritten but increasingly permanent constitution. Under such quasiconstitutional oligarchy, plural power centers might interact in an orderly fashion, with outcomes on all but the most crucial security issues not only no longer predetermined, but openly acknowledged as such. Even decisions on the "nondebatable" issues would become less and less likely to contradict the prior understandings held by those below the highest echelons.

Harsh geopolitical and economic realities will help shape the thinking of these leaders.[21] As informed Soviet officials become increasingly aware of the real limits of their abilities to keep the global promises so easily given out in previous years, the common presumption that some substantial portion of the burden must and will be shed will continue to grow. Only by disregarding their own experts will Soviet leaders of the late 1980s be free to assume that the Soviet Union can remain simultaneously ready for war in Europe, war with China, and strategic nuclear exchange with the United States — at the same time raising the standard of living of the entire Soviet populace and investing generously in several different domestic economic sectors, each of which is deemed essential to the system's future basic performance.

The choice of what part of the burden to set down may emerge in a new round of internal debate — sometimes esoteric, sometimes in plain view — as to whether the United States, China, or a potentially reunited Germany poses the strongest actual threat to Soviet national security. Such collective deliberations seem likely to be lengthy, and may feature as much backward as forward movement during long intervals. Even as it represents a cumbersome collective effort to grapple with complex, genuine issues, this style of decision may simultaneously serve a constant effort to extract as many tactical concessions as possible from as many adversaries as possible. Scrutiny by Western analysts of particular developments in this process, whether major leadership addresses or "trial balloons," may have the additional effect of further confusing the already chaotic debate over fundamental Soviet capabilities and intentions.

The decision to continue the "dirty war" in Afghanistan is likely to be high on the Andropov agenda. Soviet leaders in search of an inexpensive yet conspicuous way to signal their commitment to pragmatic policies would surely have to look no further than the record of 3 years of costly and demoralizing misadventure in that country, whose geopolitical importance may indeed be vastly overrated in many quarters. The political costs of abandoning the Babrak Karmal regime should not be underestimated, however, even in the absence of formal dissent mechanisms or public protest. The costs of the conflict itself are no doubt manageable even in hard times. In political terms, the process by which Soviet power was brought to the Central Asian portions of the czarist empire would be placed under a thick cloud following any decision to cede Afghanistan to an anti-Soviet "Basmach" regime—which might in turn spill over into broader debates about the sources of political legitimacy for such basic practices as collectivized agriculture and central industrial planning, introduced during the same historical period.

An even greater opportunity to exert diplomatic and strategic leverage may arise as Soviet efforts to stabilize and improve relations with China continue.[22] The ability of Soviet leaders to maintain their political poise in the face of a possible U.S.–Chinese strategic entente has already faced—and seemingly passed—a stringent test during the 1978–1981 period. In the context of "great triangle" diplomacy, the wisdom of reducing tension with China should be obvious to Moscow and Peking alike. But the belief that substantive improvement in Sino–Soviet relations is likely in the near term should be tempered by close attention to bilateral issues that may sooner rather than later dilute both parties' interest in impressing the United States and Japan with their freedom of diplomatic action. The Soviet Union's claim to the status of Asian power clashes with China's renewed commitment to primacy in the region at numerous points, including Mongolia, Korea, India/Pakistan, Afghanistan, and most especially Indochina. On another plane, where questions of economic and technological development are of greatest importance, both the Soviet Union and China are in sharp competition for favored access to Japanese capabilities and resources. In short, while regular communication between Moscow and Peking seems likely to be restored, the messages communicated between the two capitals will contain more references to conflict than cooperation until one or the other credibly signals that its basic goals and ambitions in the area have changed.

In search of fresh sources of internal strength, Andropov and his colleagues may conclude that the most vivid contrast to the Soviet Union's present-day economic inefficiency is the country's remarkable cultural resilience. Astute Soviet leaders have long been aware of the deep and enduring resonance that some of the traditional themes of Russina life and thought have retained more than 65 years after the end of the autocracy. If Brezhnev's successors are tempted to choose drastic means to unlock the power they seem convinced their nation possesses, their preferred key could well be Russian nationalism.

The unleashing of such a political and cultural force might primarily affect the technical and scientific intelligentsia—still wrestling with questions of East–West identity and competing values that were first raised during the nineteenth century—rather than the industrial and farm workers that make up the bulk of the Soviet populace. The appeal of such interlocking themes as Siberian development, agricultural renovation through some form of restored private incentive, disgust with corruption, and the professed need to retrieve and cherish traditional, largely rural values may not be overlooked by Soviet leaders impatient at the ossification of the party apparatus and the increasingly acknowledged inadequacy of the propaganda system.

A decision to politicize these tendencies from above—rather than await their politicization from below—would require extraordinary resolve on the part of Soviet leaders, as well as willingness to abandon symbols and procedures that have made Lenin and the Leninist party an icon for 60 years. Yet this will remain one avenue open in front of Brezhnev's successors, eager to make their decisions about where to advance and where to disengage with the active support rather than the cynical indifference of millions of educated Soviet men and women.

In foreign policy, a conscious resort to what might be called "socialist nationalism" (as distinct from the exhausted appeal to socialist internationalism) might not necessarily entail the tactics of gross expansionism or chauvinism in either Eastern Europe or Asia. Conjecture along these lines might instead center on a possible effort to replicate Russia's search for an acknowledged role in the international system of nineteenth-century Europe, with willingness both to define St. Petersburg's interests and to accommodate such definitions proferred by other members of the system. In any event, this would match already evident tactics of consciously emphasizing the "reasonable" qualities and supposed continuities of Soviet diplomatic aims, all perhaps part of a new concert of Europe.

This highly speculative presentation may or may not be of interest to readers of this essay. Of more importance over the next few years of the Andropov era will be whether similar speculative presentations in Moscow will be of interest to Soviet leaders.

U.S. POLICY AND POLITICAL CULTURE: MYTH IN SEARCH OF REALITY

The last shreds of the internationalism developed during and after World War II are now disappearing from U.S. foreign policy. In the attitudes of thousands of individuals ordinarily concerned about public affairs, this internationalism is being replaced by an essentially isolationist denial that any specific issue in dispute—whether with allies or adversaries—has any intrinsic importance. Some observers may even believe that fundamental U.S. abilities to reckon over costly foreign policy issues without primary reference to domestic political impact, never high at any point, is now falling rapidly.

Arriving along with this "neoisolationism"—perhaps an integral part of it—is a belligerent form of economic nationalism. This vociferously expressed sentiment has hobbled ongoing diplomatic efforts to negotiate orderly changes in the rules of trade and competition, as the volume of talk of future trade wars among GATT members begins to rise. Yet it has wrought little if any change in the consumer tastes and market behavior of the U.S. public, which continues to reject a multitude of domestic products on grounds of both quality and economy.

This shift reflects a basic failure on the part of virtually every institution and leader concerned with U.S. foreign policy. Specifically this is the failure to appreciate—or to convince others to appreciate—that the decline of U.S. political and economic primacy has been a secular rather than voluntary process, which cannot be reversed by "strong leadership" alone. Few members of the informed public seem willing to acknowledge that the desire to restore the international balance of power largely as it existed in 1945 is futile, and that those who promise a surefire way to accomplish this goal are selling political snake oil.

It may be a profound historical irony that domestically, the United States set an example for the world of how power can be shared yet be effectively exercised. How will the United States political system now keep faith with its members, who allow it less and less time to keep its promises, yet concentrate on exercising—and sharing—stable authority over world affairs with foreign governments that also face grave domestic challenges?

Many observers over the last several years have blamed Congress for much of this, with emphasis on the difficulties posed by the War Powers

Act and by the requirement that major arms sales abroad withstand the test of possible legislative veto. The presumption is strong that Capitol Hill was once a less contentious place—and only when strong leaders and loyal caucus members make it so again will U.S. foreign policy proceed unhampered. Perhaps this is so, but the collapse of the private consensus on Capital Hill that once made Congressional leaders real figures of authority would have taken place even without the Vietnam War or the War Powers Act, given the demographic trends and judicial actions of the 1960s. Even for the 1980s, many would still acknowledge that the Congressional appropriation process remains the most consistent forum for conciliating already clashing interpretations of the national interest, in the end compelling all parties to acknowledge an authoritative system of priorities.[24]

The problem may not be with the legislature but with the legislators—and even more with their constituents. Both groups alike lack both opportunity and incentive to achieve fluency in dealing with even a few of the complex issues of foreign policy. There may be no other country active in international affairs whose lawmakers and taxpayers routinely possess so meager an active knowledge of history, geography, global demographics, and resource use patterns. Perhaps this is due to the fact that as Americans, we have never had to contemplate America's impact on our world.

The need to make the U.S. foreign policy system work effectively, at home and abroad, comes ahead of any specific issue, even one as potentially crucial as strategic arms talks or Middle Eastern diplomacy. The question to be answered is whether more than a select handful of U.S. leaders and officials can emerge as serious spokesmen on important issues, in the eyes both of foreign leaders and of members of the U.S. public.

Unfortunately, the likely answer becomes clear as soon as the question is rephrased slightly. The query is in essence whether a new U.S. foreign policy elite—a working elite, it should be stressed—might emerge and succeed in making policy formulation and conduct a comprehensible process of collective decision making in the eyes of other concerned parties. Crucial to any such tendency would be a consistent ability on the part of all advocates to realistically identify and analyze the goals and interests of competing advocates and institutions. But the idea of a renovated, insulated professional elite with a mandate to manage U.S. foreign relations would seem too preposterous in too many quarters to be taken seriously for more than a moment. Foreign relations seems likely to remain part-time work, at least for those responsible for the "hard decisions."

The task of rediscovering a central consensus for U.S. foreign policy will be complicated in the near future by the shifting geopolitical agenda, according to which neighbors and allies rather than known adversaries are emerging as the source of the most pressing problems.

If the question of just which foreign states constitute Washington's main preoccupation could be answered by enumerating major controversies that monopolize the front pages, the response might begin with the Soviet Union, then include such governments as Israel, Cuba, Libya, Iran, China, and South Africa. Issues that revolve around these countries consistently provoke the emotions, and sometimes the funds and organizing efforts, of fairly extensive segments of the public.

But if the question instead were which countries by their actions and policies have the greatest tangible impact on basic U.S. interests, related both to economics and to alliance maintenance, a quite different response would emerge. Such countries as Japan, Canada, Mexico, Nigeria, Saudi Arabia, the members of the European Community, Taiwan, and South Korea would appear. These states have multiple trading and banking arrangements with the United States that directly affect prices and trade balances that perhaps better deserve front-page treatment in an increasingly interdependent, yet more and more nationalistic community of industrial states.

The advent of industrial competition among nominal allies can hardly be called a surprise. These challenges are the price of success in prior policy efforts to stimulate development in countries like the newly industrialized states of East and Southeast Asia—efforts that were a key element of some of Washington's prior approaches to regional security. Now, however, the world outside the Soviet orbit increasingly resembles the trading floor of a major commodities exchange—at worst in panic, at best in a mess. On issues related to prices and balances among nations, it is hard to consistently close contracts that enable both partners to win.

The United States is not the only country unwilling to accept a series of costly tactical defeats in this competition. With a bitterness reserved at other times for human rights violations in the Soviet Union, Washington can (quite aptly) point to practices pursued by its major trading partners that add up to a double standard on issues of product protection and market isolation. The cost of leadership over the industrialized world seems to be going up, just when the apparent benefits are headed downward.

In terms of geopolitics, the most striking development during this period of transition may be the return of U.S. foreign policy to the North American continent. Concrete developments in both Canada and Mexico have summoned all parties interested in such homely issues as the U.S. automobile industry or the demographics of major market retailing to become more aware of the political and economic discord that exists between the United States and its immediate neighbors. The complexities of resolving disputes over fisheries or gas pricing that directly involve not only foreign ministries but multiple governmental bureaucracies, as well as numerous firms, labor unions, and private individuals, are overtaxing enough. How much worse it becomes when the realization grows that these costly problems in foreign relations will respond neither to the threat of military intervention nor to the willingness to sell arms—the two primary tools of a superpower not yet prepared for the postindustrial era. With reference to Mexico and Central America in particular, the fallout effect on U.S. immigration and labor policy exerted by the demographic and economic realities prevailing in these countries may succeed where reams of "do good" educational efforts have failed—thousands of Americans, without leaving their home communities, are now more likely to meet thousands of other citizens of the world who dream of living like Americans. Whether the meeting will be pleasant is more difficult to say.

Perhaps the most fascinating case for the rest of the 1980s will be that of U.S.–Japanese relations. Beyond the specifics of market access and bickering over levels of defense expenditures lies the question of what basis for close bilateral cooperation actually exists. The Japanese are as entitled as any sovereign people to ask what net political advantage abroad they will reap in return for acquiring a combined reputation for militarism and subservience to the United States. Are U.S. and Japanese interests so patently complementary that new procedures aimed at shared decision making on security and political issues would be redundant? Can technology be shared for defense but remain secure from industrial espionage by either side? So long as such questions remain unanswered—indeed, largely unasked, except by a small band of scholars and specialists—the clear redefinition of Japan's responsibilities and prerogatives in the alliance that must accompany any substantial increase in Japanese military capabilities will not take place.

How U.S. policy toward Japan will evolve in the absence of such an increase is an urgent, open question. As U.S. public sentiment turns increasingly toward blaming Japan for structural problems in the U.S. industrial economy—and as Japanese public sentiment turns increasingly toward blaming the U.S. tendency to blame Japan for structural problems in the world marketplace—the chances of a major disruption in U.S.–Japanese relations, though still small, nonetheless increases discernibly.

The entire question of distinguishing rival allies from absent enemies in U.S. foreign policy during the 1980s brings us full circle to the original inquiry about fundamental change in the political culture underlying U.S. foreign policy. Stated simply, meaningful consensus is lacking over virtually every issue pertaining either to potential military engagement or to likely economic conflicts along lines irrelevant to U.S.–Soviet rivalry. Rhetorical preoccupation with Soviet intentions and the necessity of matching Soviet military capabilities, while not spurious in themselves, mask this lack of consensus. In the prevailing atmosphere, partisan antagonists in U.S. domestic politics appear to have reached a strong tacit agreement that regardless of the specifics, the basic acquisition of influence in world affairs leading to tangible success and intangible prestige should always be a fast, cheap, and enjoyable process.

Some analysts may contend that the deployment of U.S. forces to help manage a truce in Lebanon may signal a new willingness to accept responsibility; but the real test may lie ahead, with presidential campaigning and actual casualties among U.S. personnel. Such a commitment of specialized military forces for a defined political goal—especially one not involving hostilities with another major state—was at one time fairly routine and not especially burdensome for the "traditional great power."

A pessimist might conclude that the "traditional great power" is truly extinct, and may never have flourished in North America in the first place. The possibility that a new pragmatism will arise in U.S. security policy and diplomacy, implemented by a well-insulated "new elite" of professional diplomats, seems small indeed. The enduring free-for-all atmosphere of U.S. political life, coupled with a narrowing rather than widening tolerance for ambiguity and compromise among domestic leaders and institutions still not coping with new scarcities, seems to promise only an acceleration of the centrifugal forces tearing at the U.S. foreign policy system.

In short, the United States seemingly cannot stop acquiring rudimentary influence abroad, based on its potential to confer benefits and penalties to every other actor in the world arena. This process of constant regeneration will paradoxically intensify during the 1980s if the United States emerges as the indispensable consumer of the world's cash crops and export merchandise, and hence the single market with greatest impact on worldwide economic conditions. No other state can reckon a strategy of its own without taking into account the likely response of the United States—even if this is in essence the likely response of the U.S. market. But when that response is increasingly perceived in political terms as either inherently unpredictable or fatally self-contradictory, U.S. policy appears to take on the characteristics of a natural hazard in world politics rather than an organizing force. Under such circumstances, U.S. foreign policy will increasingly resemble the weather—something everyone talks about, but no one does anything about.

The United States, in short, may be the first superpower in modern history to have neutralized its own influence abroad in the course of trying to maintain both democracy and prosperity at home, just when its physical capability to exert such influence was at its height. In the nuclear era, this would be a truly monumental political failure.

TOWARD THE RESUMPTION OF SOVIET–AMERICAN RELATIONS?

As the mid-1980s begin, the United States and the Soviet Union hardly seem to be members of the same system of international relations. Filling each other's rhetorical horizons, in practical terms Washington and Moscow are invisible enemies—they sense each other's presence, but avoid contact, in the form of either dialogue or conflict. The internal political, economic, and cultural changes under way in both systems may accentuate this trend—or may invite the leaders of one or both countries to adopt a new self-perception that will stimulate fresh engagements abroad.

Given the energy surges and sudden disruptions that are likely to accompany major changes in elite political culture, the specifics that will govern policies of both the United States and the Soviet Union toward mutual relations for the rest of the decade seem especially unpredictable. Whatever the uncertainties, any episodic return to the traditional belief that military power must and does translate directly into political influence abroad will aggravate existing instabilities in the relationship.

Even in their mutual isolation, both the United States and the Soviet Union have continued the dim search for influence over events abroad, mostly on a basis of "one problem, one region at a time." Both apparently continue to accept the conventional wisdom of geopolitics that somewhere a lever exists that will put the opponent at a decisive disadvantage, yet at minimum cost and casualties to the victor. In any event, such a conviction appears to be the best consolation to statesmen and commentators forced to realize that even an officially designated global superpower can no longer afford to be everywhere at once.

A more sophisticated variation on this tendency toward exclusive focus on one problem at a time in search of the final hammerlock is the tendency to argue that the opponent's internal troubles are so much worse than the protagonist's that time is squarely on the protagonist's side. Such an assumption, although it does not necessarily create a sense of security, does justify a certain degree of inactivity, supposedly temporary.

Over time, this inactivity emerges as the functional equivalent of isolationism, or "neoisolationism." A concept yet to be fully defined, neoisolationism at first glance seems to combine a retained sense of the pertinence of events taking place beyond one's frontiers with a resolve not to risk costly failures in areas where the opponent—compelled to move first—is more likely to experience defeat and frustration.

Once this neoisolationist insulation has worn thin, however, strong surges of political energy could be directed abroad, involving a heavy wager on neotraditional beliefs about the link between military capability and political influence. Could this lead to a direct Soviet–American engagement? A "Cuban missile"-style confrontation over the Middle East or even Cuba once again might trigger an actual war between the United States and the Soviet Union, implausible as that may seem; the pretext, after all, would still easily exceed the importance of an archduke's being shot. Such a crisis, on the other hand, might be a highly positive development toward the resumption of U.S.–Soviet relations, affording the most efficient path to a restored dialogue between the nuclear superpowers.[25]

An alternative would be the simultaneous rise within the policy elites of both countries of a strong new consensus in favor of vigorous negotiation in areas of mutual involvement, both functional and geographic. Clearly this would entail high political costs to be paid internally on both sides. For the U.S., the sensitive issue of apparent capitulation to Moscow—in the guise of according legitimacy to the routine exercise of power by the Soviet Union in its declared areas of interest—would have to be placed securely out of bounds from mainstream partisan political competition. For Moscow, a new round of productive exchange with the United States would probably entail an unmistakable admission of inferiority to Washington in numerous areas of common concern, at least below the level of comparative nuclear arsenals. The Middle East would be conspicuous in this regard, as might East Asia as well, if the Taiwan issue has not disrupted Sino–U.S. relations. The odds against such a "happy ending" coming about simply and quickly are clearly formidable.

Prior to some form of resumption in U.S.–Soviet relations, the principal beneficiaries would appear to be those countries that might be called the second- and third-rank powers—especially those outside Europe that seem determined to obtain extensive, technologically sophisticated military establishments of their own. So long as the Soviet Union and the United States remain prepared to transfer either funds or actual military hardware on suitable terms to these countries, their efforts will be largely unhampered during the time required to develop indigenous production capabilities for appropriate weapons and supplies. This category of states, it might be conjectured, will include those regimes most likely to focus on single issues of professedly intense national importance and equally acute regional impact. Proficiency on the part of such states at manipulating American and Soviet policies alike will constitute an important supplemental source of power. Awareness of this proficiency, on the other hand, may act as a powerful stimulus toward neoisolationism for both Washington and Moscow.

Diffusion of power and authority in international relations along these lines will pose once again the old question of whether the emergence of a single regional organizing power or the evolution of an adjustable regional balance will prove most likely to preserve stability and encourage real growth in several major world regions. If the basic political and strategic link between the United States and Israel were ever to be disrupted, for example, this question would have to be resolved on its own terms in the Middle East, with both superpowers conceivably absent from the proceedings. The success or failure of policies adopted by India after Indira Gandhi and Iran after Khomeini will also be critical indicators of likely trends, as will the evolution of Vietnam's conflicts with its neighbors. Japan too will be a major case in this regard, capable under domestic political pressure of setting a rapid pace toward regional competition, both military and nonmilitary, and equally capable of setting a vivid counterexample of assertive effort to preserve and promote regional balance.

Might a new style of international system, capable of genuine multilateral diplomacy among allies and adversaries alike, come into being? The utopian perspective on this question, however intellectually stimulating, is probably of less practical relevance than ever in an atmosphere of intensified nationalism. But what of the wish for a second chance at making a nineteenth-century-type system work, with emphasis on the elite consensus among diplomats and statesmen that often seemed to bridge national frontiers at that time. This would seemingly square with the nineteenth-century themes that sometimes appear to pervade the conflict over political fundamentals taking place in both the United States and the Soviet Union. Yet the availability of "small wars" to adjust the balance—preferably in remote regions like Afghanistan—is now open to question, however; even the destructiveness of conventional weaponry may be enough to demolish the proposition of war as an occasional but necessary expedient in the affairs of an acknowledged, basically secure great power.

In any case, even if an objective concert of interests among statesmen no longer preoccupied with ideological conflict were to emerge during the 1980s, an essential political and cultural building block of the Victorian era would appear to be irretrievably missing. The elites responsible for making such a global consensus match up against domestic political understandings in fact hold their positions of quasilegitimate authority only tenuously, and may deem it necessary for self-preservation at any time to mobilize their constituents against foreign enemies, real or imagined, rather than risk forfeiting their mandate.

This essay may have conveyed a sense of dismay or dissatisfaction with regard to the present state of political affairs encountered at both ends of the Washington–Moscow hotline. It is all the more ironic then, to come to the still tentative conclusion that the present deadlock–involving enemies that are mercifully invisible–may indeed be preferable to whatever will come next, when the gladiators compose themselves sufficiently to return to a much-changed arena.

The essence of Soviet–American relations will remain the mutual decision as to whether nuclear deterrence will prosper or fail; as a relationship of influence, deterrence is inherently unlikely to remain static. The enduring importance of the elusive problem of monitoring changes in elite political culture for both contestants reflects a fundamental reality of deterrence–it is a relationship sustained not by the size of two arsenals but by psychological processes involving both individuals and complex groups. Many observers have noted the significance of mutual perception in this regard. The trends at work during the current period strongly suggest that the phenomenon of elite self-perception, as a source of values capable of governing political decisions under stress, should receive comparable analytic scrutiny.

NOTES

1. Of the many basic sources from which more detailed, better documented versions of events can be obtained, perhaps the most serviceable would be annual register published by the Council on Foreign Relations, entitled *American Foreign Relations: A Documentary Record*.

2. For a discussion of the overlapping concepts of political culture, ideology, and political organization that pertains primarily but not only to the Soviet Union, see Alexander Dallin, "The Domestic Sources of Soviet Foreign Policy," in Seweryn Bialer (ed.), *The Domestic Context of Soviet Foreign Policy* (Boulder, Colo.: Westview Press, 1981), pp. 355–360, 396–397n. See also Bialer's own chapter in the same volume for a related discussion of the ideology and political culture, taken from Clifford Geertz, "Ideology as a Cultural System," in David Apter (ed.), *Ideology and Discontent* (New York: Free Press, 1964).

3. For an early but definitive examination of the organizational task of leadership, see Phillip Selznick, *Leadership in Administration* (New York: Harper & Row, 1957).

4. See Richard P. Stebbins and Elaine P. Adams, *American Foreign Relations 1975: A Documentary Record* (New York: New York Univ. Press, 1977), especially pp. 24–29 on the Trade Act of 1974 and pp. 603–618 on Angola. For a good glimpse at the debate then in progress, see also U.S. Senate, Committee on Foreign Relations, *Detente: Hearings Before the Committee on Foreign Relations* (Washington: U.S. Govt. Printing Office, 1975), especially the testimony of then Secretary of State Henry Kissinger and such leading Soviet affairs experts as Marshall D. Shulman and Herbert S. Dinerstein.

5. See the text and interpretation of President Carter's major address at the U.S. Naval Academy in June 1978 in Elaine P. Adams (ed.), *American Foreign Relations 1978: A Documentary Record* (New York: New York Univ. Press, 1979), pp. 18–32, 204–210.

6. For the text of the addresses and documents presented in Vienna, see "Text of the Joint Communique by Carter and Brezhnev," along with other materials, *New York Times*, June 19, 1979.

7. See Bialer's essay appearing in the *Washington Post*, January 16, 1980. For a provocative, more recent account of the record and outlook for U.S.–Soviet relations, see Seweryn Bialer and Joan Afferica, "Reagan and Russia," *Foreign Affairs*, Vol. 61, No. 2, pp. 249–271. For a countering view in the same issue, see Louis J. Walinsky, "Coherent Defense Strategy: The Case for Economic Denial," pp. 272–291.

8. Press conference of January 29, 1981, excerpts reprinted in *Department of State Bulletin*, Vol. 81, No. 2048 (March 1981), p. 12.

9. *Ibid.* p. 13.

10. *Ibid.*

11. From text of address reprinted in *Department of State Bulletin*, Vol. 82, No. 2064 (July 1982), p. 27.

12. *Ibid.*

13. *Ibid.* p. 28.

14. *Ibid.* p. 25.

15. From text of address reprinted in *Department of State Bulletin*, Vol. 82, No. 2063 (June 1982), p. 36.

16. *Ibid.* p. 35.

17. Of the many useful writings to appear following Brezhnev's death, see especially Seweryn Bialer, "The Andropov Succession," *New York Review of Books*, Vol. XXX, No. 1 (February 3, 1983), pp. 26–30.

18. See, for example, T. H. Rigby, "The Soviet Leadership: Towards a Self-Stabilizing Oligarchy?," *Soviet Studies*, No. 22 (1967), p. 167. For an introductory text in Soviet politics and government that takes a related approach, see Darrel P. Hammer, *USSR: The Politics of Oligarchy* (Hinsdale, Ill.: Dryden Press, 1974).

19. This could be considered a vote for a closer look at the organizational process model laid out by Graham Allison in *Essence of Decision* (Boston: Little, Brown, 1970) for pertinence to the Soviet policy making process. For a more sustained argument in favor of this, see Arnold L. Horelick, A. Ross Johnson, and John D. Steinbruner, *The Study of Soviet Foreign Policy: A Review of Decision Theory-Related Approaches (R-1334)* (Santa Monica, Calif.: The Rand Corporation, 1973).

20. For an unexpected confirmation of the size of the Soviet polity, see the official estimate that for nuclear targetting purposes, 110,000 office holders are to be considered key to the Soviet government and party apparatus, conveyed in Desmond Ball, "U.S. Strategic Forces: How Would They Be Used?," *International Security*, Vol. 7, No. 3 (Winter 1982/1983), p. 55.

21. For an articulate discussion of political–military basics of Soviet policy, see David Holloway, "Military Power and Political Purpose in Soviet Policy," in *Daedalus*, Vol. 109, No. 4 (Fall 1980), pp. 13–30. In the same issue, see Richard Pipes, "Militarism and the Soviet State," pp. 1–12.

22. For recent surveys of Soviet policy in Asia, with focus on Sino–Soviet relations, see Herbert J. Ellison (ed.), *The Sino–Soviet Conflict: A Global Perspective* (Seattle: University of Washington Press, 1982), and Donald S. Zagoria (ed.), *Soviet Policy in Asia* (New Haven Conn.: Yale Univ. Press, 1982).

23. For linked perspectives on the basic problems and prospects of U.S. foreign policy, see Louis J. Halle, *Dream and Reality: Aspects of American Foreign Policy* (New York: Harper & Row, 1974), and David P. Calleo, *The Imperious Economy* (Cambridge: Harvard Univ. Press, 1981).

24. For a recent essay surveying the issues in legislative–executive relations over foreign

policy problems, see Warren Christopher, "Ceasefire Among the Branches: A Compact in Foreign Affairs," in *Foreign Affairs*, Vol. 60, No. 5 (Summer 1982), pp. 989–1006.

25. Among useful treatments of the issue of the psychological underpinning of relationships of deterrence, see Robert Jervis, "Deterrence and Perception," in *International Security*, Vol. 7, No. 3 (Winter 1982/83), pp. 3–30, as well as the same author's review essay "Deterrence Theory Revisited," in *World Politics*, Vol. 31, No. 2 (January 1979), pp. 289–324.

8

National Security and the Policy Paradox

James C. Hsiung

Throughout this book, we have been seeking an answer to the questions we have posed: Is there *policy* in the Reagan Administration's management of U.S. interests in Asia? If so, how is it *different* from the past, more especially from the Carter policy? How, or by what *strategy*, is the policy being carried out? and with what *results*? Although the chapters above look at the questions from different perspectives, some generalizations can be made with considerable congruence and coherence. There is, for example, general agreement that the Reagan Administration does seem to follow a set of goals, although the results of their implementation thus far may raise doubts as to whether the goals are consistent and achievable and, above all, whether mere postulation of goals constitutes policy per se and whether these goals are necessarily in the best interests of the United States despite their good intentions. The chapters may disagree on the extent to which the Reagan policy and strategy continue, or depart from, those pursued by previous Administrations. They agree, nevertheless, that there is no complete break from the past, only modifications (at times, tinkering), and that the differences are mostly in emphases and assumptions.[1]

GOALS AND STRATEGY

Basic Foreign Policy And Central Strategy

From the very outset of the Reagan Administration, four foreign-policy goals were postulated, and these have since been reiterated: Restraining the Soviet Union, reinvigorating U.S. alliances, strengthening ties with friends, and a more effective approach to the developing countries.[2] The first goal logically leads the other three. Strengthening Washington's alliances, forging new ties with friends (including China), and encouraging Third World nations to join the U.S. bandwagon are all ancillary to the first and foremost goal of isolating and restraining the Soviet Union. While the anti-Soviet premise has permeated the foreign policy of all post-World War II U.S. Presidents, the latest incumbent perceives the Soviet Union not only as a rival power menacing U.S. security (in the sense of nineteenth-century balance of power), but a deadly threat to the very existence of the Western capitalistic system (in an ideologically determined zero-sum game). That fear for survival has only been reinforced by Reagan's anxiety created by unrelenting Soviet advances in nuclear and conventional power.

The central strategy adopted under Reagan consists of two distinct parts: (1) Remedying the perceived nuclear and conventional imbalances ("closing the window of vulnerability"), through simultaneously increasing U.S. defense capabilities and getting the Soviets, through negotiation, to accept reductions of armaments; and (2) attempting a *global* defense that encompasses Western Europe, the Persian Gulf, the Indian and Pacific Oceans, and the Caribbean. In addition to nuclear deterrence, this plan relies on the building of a "three-ocean navy" and embraces a strategy of "horizontal escalation" and "sustainability." (See Chapter 1.)

Reagan's Asian Policy and Regional Strategy

In Asia as elswhere, the Administration asserts that it wants to reverse a decade of U.S. decline in power and Carter's vacillation, curb Soviet expansion, reassert U.S. leadership, and ensure access to vital resources. To achieve these objectives in the region, as Normal Levin points out, five measures are essential: (1) a renewed and unequivocal commitment to U.S. allies as well as help to victims of Soviet "aggression"; (2) dismantling of restraints on arms sales abroad (imposed by Carter on human rights grounds) to deter "aggression"; (3) a greater emphasis on defense considerations and military capabilities, more globally coordinated than before; (4) building a "strategic consensus" in the region to contain Soviet influence; and (5) significant U.S. military expansion into the Persian Gulf and Southwest Asia.

Accordingly, South Korea's security has gained added significance, because it is a very crucial piece on the United States's strategic map, not just as a gateway to Japan, as before. In the strongest showing of support in 12 years, the Reagan Administration has reaffirmed the U.S. nuclear umbrella for this "pivotal" Asian country, and has initiated a tremendous increase in the sale of modern weapons and greater cooperation in the manufacturing of weapons in South Korea, some to be exported to other countries. Even more than before, Japan is considered a "full partner" in defense, trade, and world-order matters, and is to be encouraged to assume sea and air patrols west of Guam and north of the Philippines. Thus far, relations with South Korea have considerably improved, and the uncertainty created by Carter's contemplated (but later suspended) plan to withdraw U.S. troops from Korea has dissipated under Reagan's assurances, backed by both a military assistance program to increase Seoul's immediate combat readiness and substantial economic concessions. However, inherent tensions with Japan, arising from the United States's $20 billion trade deficit with that nation and Tokyo's resistance to U.S. pressures for a greater "burden sharing" of defense expenditures, are continuing, as under previous Administrations.

Burgeoning domestic protectionist sentiments in the United States against the onslaught of Japanese goods are indicative of the strained relations, which are reaching a level unprecedented since 1945. The visit in Washington by Yasuhiro Nakasone, the new Japanese Premier, in January 1983, failed to produce a breakthrough on trade, although he expressed his government's willingness to "share responsibilities" in maintaining peace through military defense. Without making any new promises, Mr. Nakasone said he was willing to put into effect an existing 5-year plan to increase and extend Japan's military defense (albeit still falling short of U.S. expectations) deeper into its airspace and farther into the sea lanes that separate it from the Soviet Union.[3]

In its application to China, the Reagan plan includes a commitment to sell it lethal weapons, reversing the Carter policy of selling only "dual purpose" (civilian and military) technology and equipment. Announced during Secretary of State Haig's June 1981 visit to Peking, the lifting of the previous U.S. weapons ban was meant to encourage the Chinese to accept a "strategic consensus" with regard to the Soviet Union. As Robert G. Sutter argues, this was one way of putting pressure on the Soviet Union and of drawing China closer to the United States, as well as pressuring Japan into accepting an expanded military burden. However, due to Chinese foot-dragging and new hitches in Sino–U.S relations—in which the Taiwan impasse played only a part—the U.S. offer of arms sales and an implied quasi-alliance was not picked up, as expected.

The Chinese aloofness, apparently, has survived the temporary resolution of the Taiwan arms-sales impasse worked out in the new U.S.–Chinese communiqué of August 17, 1982. In it, Reagan acceded to the Chinese demand that the United States ultimately terminate its arms sales to Taiwan, over which Peking claims sovereignty in conflict with the Kuomintang government ruling the island. In assessing the emergence of a new "independent" Chinese foreign policy, which seeks to place a relative equidistance with Washington and Moscow, Carol Lee Hamrin discovers that the new posture of the Deng Xiaoping leadership is the result of a relatively long review, from 1979 to 1982, and is not a temporary coup. The most recent trade difficulty, one may add, is probably an indicator of Peking's "independent" posture: In rapid-fire succession, Peking promptly banned further imports of cotton, soybeans, and chemical fibers from the United States, in retaliation for the latter's curbs on Chinese textile imports imposed almost immediately after both sides had failed to reach a new agreement on Chinese textile sales in January 1983.[4]

The Chinese, in Hamrin's study, are found to harbor a resurgent suspicion of Washington and willing to inject a new realism in dealing with Moscow. While these shifts are no doubt linked to the domestic political climate, they are also clearly related to China's desire to improve its standing and influence in the third world. The latter is now receiving renewed priority on Peking's foreign-policy agenda, as can be witnessed in Premier Zhao Zhiyang's 1-month trip to 11 African countries in late 1982–early 1983. The new Chinese policy of evenhandedness with the two superpowers is as much designed to avoid becoming a "pawn" to be used by either as to explore the possibility of alternative Soviet bloc sources of input to aid Chinese development. Behind these shifts the Deng leadership appears to have solid domestic support, especially following the 12th Party Congress and the 5th Session of the 5th National People's Congress in the latter half of 1982. Paradoxically, Hamrin concludes, it is the unswerving U.S. commitment under Reagan to the cause of containing the Soviet Union that has made the Chinese perceive a decline of Soviet threats to them, which in turn has paved the way for the shifts in their foreign policy of seeking improved relations with Moscow, while putting U.S.–China relations on hold.

In Southeast Asia, the Administration's policy objectives, according to John W. Garver, are to maintain the socio-economic and political stability of the ASEAN countries (Thailand, the Philippines, Singapore, Malaysia, and Indonesia), to constrain Vietnam's regional expansionism, and to curb the growing Soviet influence in the region. Although continuing some of the policies from the Ford and Carter years, such as strengthening ASEAN, the Reagan policy is based on the premise that the steady decline of U.S. influence since the fall of Saigon in 1975 has encouraged and facilitated the growth of Soviet influence in the area. Support for ASEAN is one component of the Reagan policy toward the Asia–Pacific basin, where it is hoped that a democratic (*qua* capitalist) "Pacific Community" will emerge during the latter part of the century. Military aid to Thailand, which by 1982 increased by 39% over the Carter allotment, was thus more than just a reaction to the Vietnamese invasion of Kampuchea and its attendant threat to the Thais. It is part of the larger Reagan design. So is the much enhanced military assistance to Indonesia, up by 50% in 1982. The playing down of human rights has also produced noticeable effects in U.S.–Philippine relations. Besides human rights, the Reagan government has found a few new weapons with which to tighten the vise on Vietnam and to justify continuing U.S. hostility and nonrecognition, including attacks on Hanoi's alleged use of biological weapons. In conjunction with China and ASEAN nations, Washington has worked through the United Nations to bring pressures on Vietnam and to end all developmental aid to Vietnam and Vietnam-occupied Kampuchea, except solely for humanitarian purposes to the latter. It has also supported ASEAN efforts at forming a united front on Kampuchea. The policy is one of isolation and pressure to force Hanoi to withdraw troops from Kampuchea.

Nevertheless, Garver finds, the Reagan policy shares one thing in common with that of Carter, which is the abandonment of the traditional U.S. belief that Chinese hegemony over Indochina would pose unacceptable threats to U.S. interests in Southeast Asia. Like its Democratic predecessor, the Reagan Administration is willing to accept Indochina as a Chinese "sphere of interest" and the possibility of an enhanced Chinese standing in Southeast Asia, in the event Vietnam should choose or be pressured to end its romance with the Soviets. Perhaps more than its predecessor, it is lending warm support for the Chinese plan for resolving the Kampuchean question, the installation of a tripartite coalition supported by Peking to replace the Hanoi-backed puppet regime.

As with the Carter Administration, the constraints inherent in playing the China card with regard to Indochina have deprived the Reagan government of a freer hand in dealing with Hanoi directly. Partly out of deference for Chinese sensibilities, normalization of U.S. relations with the Democratic Socialist Vietnam continues to be an evasive object. In the absence of contacts, the United States continues to have to rely on indirect instrumentalities, such as the United Nations and Peking, to have anything to do with Hanoi. It continues to suffer from a lack of any leverage in dealing with the Vietnamese. As long as Sino–Vietnamese tensions remain at the present level, Hanoi has no prospect of freeing itself from its dependency on Moscow, which is known to be footing the Vietnamese operation in Kampuchea on the order of $5 million a day. In return, the Soviet military presence in Indochina has been expanding without relent. This, paradoxically, runs counter to President Reagan's foremost Asian policy objective of containing Soviet influence in the region.

The drive to establish a strategic consensus against the Soviet Union was particularly evident in the Middle East, in the first year of the Reagan government. It was as though the momentum of the peace process started under Carter with Camp David had come to a halt. State Department efforts under Secretary Haig were concentrated on sensitizing moderate Arab states in the region to the danger of Soviet expansionism and rallying their, and Israeli, support for U.S. efforts to keep out the Soviets and to assure the oil flow from the Persian Gulf. If the Arab–Israeli conflict had dominated the Middle East policy of the United States since 1948, the Reagan Administration has developed a parallel nexus, which makes Saudi Arabia the Gulf region's main supporter of the U.S. campaign to contain the Soviet Union. Soviet influence has been considerable in Syria, Libya, Iraq, and South Yemen, among others. Although begun under Carter, the Saudi nexus has been given a fuller play since 1981. The new program has provided military sales to boost the Saudi Air Force capabilities, including the sale of five E-3A AWACS (airborne warning and control system) aircraft, at a total cost of $5.8 billion. It also envisages closer military ties with Pakistan, Turkey, Morocco, Egypt, Sudan, Somalia, Jordan, and Oman. Together with Saudi Arabia, the region's cornerstone, these Arab nations constitute a new coalition that parallels the traditional U.S.–Israeli alliance.

The Lebanese invasion by Israeli forces in the summer of 1982, Winberg Chai points out, opened new opportunities for the Reagan government. Until the outbreak of the Lebanon–Syrian war of April 1981, when the Syrian SAM (surface-to-air) missiles installed on Lebanese soil threatened a direct Syrian–Israeli confrontation, the President did not seem to show much interest in the peace process begun at Camp David in 1977. Ever since then, however, he became fully aware of the volatile situation in the Middle East. The 1982 Israeli invasion brought Reagan back into what Chai calls "the mainstream of U.S. policy" fashioned by Nixon, Ford, and Carter. The crisis moved President Reagan from being a passive mediator to a full participant, not just in resolving the Lebanese occupation, but in finding a durable Middle East settlement. Calling for a fresh start going beyond Israeli Prime Minister Begin's narrow definition of the Camp David accords, Reagan on September 1, 1982, the day the PLO withdrawal from West Beirut was completed, proposed a package for Middle East peace. The following principles were endorsed in the package: (1) full autonomy for the Palestinians on the West Bank and in Gaza, and Palestinian self-government in association with Jordan; (2) a 5-year transition, as outlined at Camp David; (3) a freeze of settlements by Israelis in the occupied areas; (4) opposition to an independent Palestinian state in Israeli-occupied areas, and opposition to Israeli annexation or permanent control of these areas; (5) Jerusalem to remain undivided, but with its final status to be decided through negotiation; and (6) acceptance of United Nations Resolution 242.

This U.N. resolution, which was accepted along with U.N. Resolution 338 in the Camp David accords as the basis for a "just, comprehensive,...durable" settlement, provided for withdrawal of Israeli armed forces from occupied territories, and termination of all claims or states of belligerency and respect for the sovereignty of every state in the area and its right to live in peace within secure and recognized boundaries. In other words, the dictum "peace for territory" was once again reiterated by the United States under the Reagan plan.[5]

The Reagan call for the return of Israeli-occupied territories, though anticipated by the Camp David accords, was considered by moderate Arab states as an important step forward. As such, it was greeted positively in the form of the Fez Declaration adopted by the Arab League at the summit meeting of September 1982, held in Morocco. However, unless the President is able to push Menachem Begin into relenting on his stubborn, professed policy of annexing the occupied territories, at the pain of losing all U.S. economic and military assistance, little progress is expected to come out of the Reagan initiative beyond the much-troubled negotiated withdrawal of foreign troops from Lebanon.[6]

A REVIEW OF THE REAGAN RECORD

In summary, the Reagan record in Asia thus far has (1) strengthened ties with South Korea, whose security is dependent on continuing U.S. military presence in the peninsula; (2) not made much headway in resolving the huge trade deficits and the defense burden-sharing problem with Japan; (3) suffered temporary setbacks in Sino–U.S. relations, as China, encouraged by Reagan's commitment against Soviet expansionism, is seeking to maintain an equidistance with Moscow and Washington; (4) continued the nonpolicy, inherited from Presidents Nixon, Ford and Carter, toward Vietnam and Indochina, largely because of the lingering "Vietnam blues" in the United States and because of the constraints inherent in the China-card playing; (5) broken no new ground in relations with the ASEAN nations, although continuing lip service is paid to their importance to U.S. interests; and (6) rekindled the peace process left off by Carter with Camp David, but added a new Saudi scenario. On the last score, Reagan has yet to show resolve and ingenuity on how to break Begin's intransigence and induce him to accept the Reagan package unveiled in September 1982.

At best, this is a checkered record. To the above have to be added the not too spectacular turns of events in other parts of the world, given the Reaganite emphasis on a global view of security. Despite the Administration's early attention to the Caribbean region, the situation in El Salvador,

Guatemala, and Honduras, among others, has not changed much from the Carter years. Under the Caribbean Basin Initiative, announced with much fanfare by President Reagan before the Organization of American States, on February 24, 1982, El Salvador was the recipient of close to 50% of the $348 million in aid proposed for the region. The war in El Salvador is continuing, after the elections, and is spreading to Honduras, except that it makes fewer headlines.[7]

In the Middle East, the Soviets were said to be active again under Andropov, ready to offer a new generation of weapons systems to Syria, send in more military advisers, and strengthen ties with the PLO and radical Arab nations.[8] In Africa, there was some progress toward a settlement on Namibia that would bring about Cuban withdrawal from Angola, but it was blocked again by South Africa, which in turn was supported by Reagan. In Poland, martial law was ended in name as of late 1982, but the Solidarity movement was banned by General Jaruzelski's government. The gas pipeline embargo against the Soviet Union, imposed by Reagan because of Polish martial law and Soviet occupation of Afghanistan, had to be lifted because of opposition and defiance by NATO members.

In Western Europe, the "zero option" was losing support. The Soviets were mounting a concerted campaign to appeal to the grass-roots nuclear-free movements in Western Europe, over the heads of their governments. The election of Chancellor Helmut Kohl's government in Bonn, though heart-warming to the Reagan Administration because of his conservative bent and greater sympathy for nuclear armament, has not changed the fact that West Germany remains the home of earnest supporters for the nuclear freeze movement. Even the government of Margaret Thatcher, who has supported President Reagan's foreign policy more completely than any other European political leader, was in early 1983 beginning to speak openly about the possibility of a short-term disarmament agreement with Moscow that would meet the Soviets halfway but fall short of Reagan's zero–zero goals.[9] To arrest the Western European trend toward compromise—which might result in a total defeat of the Reagan call for Soviet dismantling of their existing SS-20 and other missiles in return for the cancellation of the NATO plan to install U.S. Pershing II and cruise missiles in Europe—President Reagan on January 19, 1983, hastily announced the formation of a Cabinet-level committee, headed by William P. Clark, his national security adviser, to promote Washington's arms-control policies abroad. Following closely the dismissal of Eugene Rostow as the head of the Arms Control and Disarmament Agency, the move was equally calculated to counter the image abroad of a government in disarray.[10]

The Reagan Administration has emphasized from the start that its success in foreign policy and national security depends on "rearming

America." On that premise, it unfurled in early 1982 a 5-year $1.6 trillion defense budget and insisted that despite the mammoth federal deficits, the integrity of its defense budget was sacrosanct.* However, after prolonged wrangling, and amidst increasing Congressional criticisms of the Administration's across-the-board weapons "wish list" without clear priority and without an apparent, coherent strategy,[11] Defense Secretary Weinberger in early 1983 was obliged to accept an $8 billion cut in the defense budget for the 1984 fiscal year. Regardless of what will happen to the rest of the ambitious 5-year plan,[12] with all the insistence on defense budget hikes and the President's open admission to Soviet nuclear superiority, the chickens seem to have come home to roost. The defeat of the Reagan defense budget would add one more evidence for the lingering impression, rightly or wrongly, that he has no coherent policy.[13]

Geostrategic Approach and Reliance on Preponderance

The current national-security program stems from the application of a geostrategic approach to security management. Geostrategic concerns about the strife between the Heartland power of the Soviet Union and the maritime alliance led by the United States for the control, or denial of control, of the Rimlands of Eurasia–Africa[14] may explain the rationale of the Reaganite global defense plan, which adopts a panoramic view of security interests. While this global geostrategic game plan is a continuation of official U.S. policy since World War II, the extension of the sphere to be protected from the reach of Heartland power from the Rimlands to the oceans, including the Indian and Pacific Oceans, is a natural addendum caused by the maritime push of the Soviet Union. The dedication with which the Administration is seeking to "rearm America" contains another traditional U.S. conservative strand, which is to fall back on the country's own preponderant power (as distinct from playing balance of power) in the management of national security.[15]

Supply-side Deterrence

The Reagan approach also contains a similar logic to supply-side economics, in that supplying more weapons into the defense equation is believed to produce greater security. Deterrence is believed to depend on

---

* The defense budget was revised in January 1983 to total $1.76 trillion for 1984–1988. The $280.5 billion defense budget proposed would be over 35% of the total $848 billion outlays in the budget the President sent to Congress for fiscal year 1984.

the size of U.S. nuclear and conventional forces, from the MX and Trident II missiles, the B-1 bomber, to the M-1 tank, backed by a declared resolve to resort to "first use." Traditional deterrence, by contrast, has always been premised on the possession of a credible second-strike capability. Its supporting theory is that if the Soviets know that they cannot get away with a preemptive strike because enough of our intercontinental missile force will survive and that we have the determination to retaliate, then they will be deterred from trying in the first place for fear of sure and costly consequences. *Deterrence*, in the usual strategic thinking, is associated with more negative goals, mainly of preventing (hence, deterring) the enemy from embarking on a course considered undesirable by us. As such, it is different from *compellence*, which seeks more positive goals, to force the enemy to do certain things desired by us against his wishes.[16]

Supply-side deterrence, however, appears to aim at confronting the adversary with such an overwhelming military might that he will be overawed (compelled) to behave. Hence, the Reagan team believes that more U.S. missiles, rather than fewer Soviet missiles (through negotiated reductions), spell more security. While the goal of deterring a nuclear conflagration is the same, the assumption is different. In supply-side deterrence, there is a confusion between deterrence and compellence. It seeks to *compel* the enemy to be *deterred*, and its means of achieving the goal is to increase our might to overwhelm the enemy, without regard for the danger of destabilization due to mutual escalation. Where conventional deterrence banks on the enemy's self-interest (fear of a costly retaliation), supply-side deterrence flaunts tougher military muscles and a desire to flex them. Its stomach for more military hardware allows no appreciation for "overkill," or why we need more and more if we already have enough to kill the other side several times over. In the present nuclear balance, only 1% of either side's arsenal is enough to wipe out 75 metropolitan centers of the other side.

Furthermore, the present Washington policy of ostentatiously extending the U.S. nuclear umbrella to allies, in NATO and elsewhere (Korea, for example), requires a questionable assumption that the United States will be ready to risk its own survival to protect its allies when they are hit by a Soviet nuclear attack. It suspends the reality of today's nuclear balance, in which the United States no longer enjoys either nuclear monopoly or superiority, as it did before.

Traditional deterrence theory posits a rational actor in its reading of the enemy's calculations. Hence, its conceptual pillar is MAD, or mutual assured destruction, which appeals to the adversary's self-interest, or fear of fatal retaliation. The supply-side counterpart, however, goes by a split assumption about the enemy: It assumes, on the one hand, that the enemy is irrational enough to be tempted to start a preemptive strike, but, on the other hand, that he is rational enough to be scared by our display of enhanced force, in fact so rational at this point that he will not be driven into another spiral of the arms race.

Whereas traditional deterrence based on a credible second strike emphasizes deescalation and plays down first use, supply-side deterrance is willing to risk escalation on the chance that the enemy will be caught bluffing. The conceptual pillar for the latter, therefore, is known as NUTS, or nuclear utilization theories. Quite typically, the current belief in Washington is that there can be a "limited nuclear war." Conventional thinking, of course, rejects such possibility because of the inherent escalation problem. In game-theoretic language, traditional deterrence follows the paradigm of the "game of the chicken," or double cooperation (CC), while the supply-side variant projects a macho-like courage and willingness to accept double defection (DD) typical of the "prisoner's dilemma" game.[17] In short, supply-side deterrence has a higher chance of a nuclear outbreak, contrary to what its adherents may believe.

Moreover, as has been pointed out elsewhere, the MX missile touted by the Reagan team could be a magnet to attract a Soviet first strike. Both because of its planned survivability in the wake of a Soviet strike and its silo-penetrating capability, the deployment of the MX missile will only compel the Soviets (1) to *empty* their remaining missiles right after their first strike and pending the arrival of U.S. retaliatory fire, and (2) to assume that U.S. decision makers, knowing ahead of time the futility of their second strike (because of the *empty* Soviet silos), actually mean first strike in their public talks about deterrence. This assumption, therefore, could tempt the Soviets to launch a defensive first strike, that is, to defend themselves from a presumed deceptive U.S. first strike, something they would not do in the absence of the MX missile on the U.S. side.[18] This is a concrete illustration of why supply-side deterrence may be self-defeating. What applies to the MX can be said of other measures that offer one side more than it needs for self-defense (second strike) against the other side, such as the ABM system. Despite its seeming attractiveness, the President's latest proposal (of March 23, 1983) of developing laser killer technology ("Star War") which will intercept and destroy enemy attacking missiles in space would have a similar self-defeating problem. It would tempt or even force the enemy to launch a preemptive strike before time$-_n$ (now or never) when the United States achieves fool-proof protection with the new system before the enemy does.

Relinquishing the Initiative

One distinct result of Washington's excessive Soviet fixation is that policy priorities are pegged on Soviet moves around the globe, to the point of forfeiting its own initiative. During the initial months of 1981, the State Department under Secretary Haig had its eyes trained on the communist (Soviet–Cuban) interference in El Salvador and the Caribbean.[19] The Persian Gulf and Southwest Asia (Afghanistan and, hence, Pakistan) came next on the list, due to the salience of the Soviet presence or proximity. From the latter half of 1981 on, however, the attention was gradually shifted to Europe, because Soviet foreign-policy priorities and concerns were very much centered on Europe. Two sets of prominent issues then caught on with Washington: nuclear weapons and arms control; and Poland.[20] Did the situation in the Caribbean really change, or was it merely that the Soviets were able to cause a change in U.S. perception?

In Asia, in general, the Soviet fixation is responsible for drawing U.S. immediate attention to Northeast Asia, because it is contiguous to the Soviet Far East air and naval bases. Maritime Southeast Asia, including the ASEAN countries, and the links between Southeast Asia and Northeast Asia have lower priority, because here the threat is less pronounced. Despite the importance of the control of the Taiwan Strait, for example, which is the necessary channel that connects the Sea of Japan and the South China Sea, Washington's strategic concerns seem to be directed at two separate segments: (1) the chunk of water stretching from the Berring and Okhotsk Seas to the Sea of Japan and the South China Sea, to the north; and (2) the links between the South China Sea, the west Pacific, and, through the Malaccan and the Indonesian straits, to the Indian Ocean, to the south.[21]

The segmentation follows the two separate operating patterns of the Soviet Pacific Fleet, one based in Vladivostok in the north (plus the Okhotsk Sea SLBM haven), and the other out of the Cam Ranh Bay, in Vietnam. As the Soviet ships do not sail through the Taiwan Strait very often, the control of that strait, it logically follows, does not occupy a top place on the United States' strategic priority list. With all his personal concerns for Taiwan, President Reagan seems more conscious of his friendship toward its people than of the island's strategic value to U.S. security. As a matter of fact, 1.5 miles off the east coast of Taiwan, between Hualien and Su-ao, the seabed sinks abruptly at a 45° angle to a depth of 4 to 5 miles, providing a perfect submarine sanctuary from which SLBMs can be launched against Soviet territories. Who controls the Taiwan Strait in wartime, therefore, would wield a great military advantage.[22] From the perspective of U.S. security interests, the Sino–U.S. wrangle over Taiwan, including the troublesome arms-sales issue, should be seen in a totally different light from that of the emotional ties by which Reagan has been identified ever since the 1980 campaign. Ironically, it would have to take more Soviet activities around and through the Taiwan Strait before it would be thrust into U.S. strategic consciousness.

Both the Soviet Union and the United States, notes Paul Borsuk, contend for the status of an Asian power, and both vie for the support of China. Both, along with China, also vie for access to Japan's advanced technology, albeit in varying degrees. The competition, which has found the two superpowers on a collision course, has paradoxically caused what Bursuk calls "a disappearance of relations" between them. Following the end of detente, the brief experiment with building a higher degree of mutual dependency (hence, mutual vulnerability) under Nixon and Kissinger, the two countries now seem to have almost no substantive, meaningful relations other than what boils down to an ever-escalating contest in missile launchers, warheads, strategic bombers, nuclear-armed submarines, etc. The eight official groups set up between the two sides, under the Carter Administration, to work on a treaty to bar satellite warfare, ban nuclear weapons in the Indian Ocean, or develop a comprehensive underground test ban have all ceased to function. Eleven commissions set up a decade ago to work out U.S.–Soviet exchanges in the fields of science and technology, seismic research, health research, and developments in space have all been suspended.* In essence, the U.S. obsession with the Soviet threat has resulted in the "disappearance" of bilateral relations. By extension, U.S. relations with allies have been reduced to a popularity contest, to turn them against Moscow on issues at stake. For example, from late 1982 on, Washington had to contend with Moscow in winning Western European support for Reagan's "zero option" over Andropov's counter-proposal.[23] The same anti-Soviet obsession has caused other policy constraints and disadvantages, which are further discussed below.

Other Policy Dilemmas

The overzealous effort to build a "strategic consensus," or a common alignment to curb Soviet influence, has, in effect, lent extra leverage to U.S. allies and friends in their dealings with us, as the preceding chapters indicate. Preoccupation with the Soviet problem may either lead to a neglect of other equally important interests, or at times produce counterproductive results.

• Japan, for example, knows that the United States needs her help in maintaining surveillance on the activities of Soviet Pacific Fleet ships in peace time and blocking their exit from Vladivostok through the narrow straits in the Sea of Japan in wartime. Thus, she is being urged to be a "full partner," in the true sense of the term, and more particularly

---

* Hedrick Smith, "U.S.–Soviet Tension Believed to Bring Ties to Low Point," *New York Times*, May 24, 1983, p. A-1.

in defense. But, that also confers greater independence on Japan in her foreign policy. Anxiously pushing Tokyo toward greater defense responsibilities does not leave Washington much leverage in getting Japanese concessions on imports of beef, citrus fruits, and other items to ease the chronic U.S. trade deficits. As long as these U.S. anti-Soviet anxieties exist, the Japanese know that they only need to give in occasionally on slight increases in defense expenditures, just to give the impression of relenting. These slight defense budget adjustments will cost Japan very little, but are sure to further boost her importance as an ally. At the same time, Tokyo knows perfectly well that there is little Washington can do to force Japan to equalize the bilateral trade. Prime Minister Nakasone's response in Washington, noted before, is a typical example. As long as Japan can give the impression that she is not as anxious about the Soviet problem as is Uncle Sam, but can be "won over" by the latter, her edge will continue.

• Greater U.S. commitments to South Korea, while instrumental in bringing about better relations with Seoul, have to contend with two real or potential consequences. One is the likelihood of a destabilization of the situation in the peninsula, as North Korea will surely try to upgrade its own defenses to match the increased South Korean war capabilities due to the sizeable U.S. arms sales and military cooperation. The other is the unlikelihood of an improvement in the human rights record in South Korea, which is now a secondary issue to Washington's overriding concerns about curbing Soviet influence. A more serious dilemma is the possibility, although remote thus far, that the heightened tensions in the peninsula, created by the massive infusion of U.S. military aid and defense buildup in South Korea, may run so serious as to impel the North Koreans to "return" to the Soviet arms in the event of a prior or expected falling out with Peking.

• We have already noted that partly because of the unmistakable Reagan brand of commitment to the containment of the Soviet Union, Peking now feels secure enough to relax its erstwhile militancy and intransigence in its relations with Moscow. Moreover, due to Washington's unwillingness to offend Chinese sensibilities, the continuing U.S. nonchalance toward Indochina, along with Sino–Vietnamese hostility, has allowed the Soviets to expand their military presence in the area. These are unintended consequences that frustrate Washington's goals of constraining the Soviet power and perpetuating the Sino–Soviet split.

• Because of the President's strong aversion for the Soviet Union, there is a disproportionate representation, in the foreign policy making process, of the high-strung views of certain hard-line Sovietologists. In Sutter's chapter, for example, it was noted that while some hard-line Soviet specialists supported arms sales to Peking, viewing them as necessary to face the growing Soviet power, arms controllers (i.e., generalists) considered such moves as detrimental to U.S. interests, because they would lessen the trust between the two superpowers. In a hair-trigger situation, known as the "balance of terror," mutual trust is the only way to set a deescalation process in motion, diverting the world away from a course of total destruction. The hard-line Soviet specialists, whose views were closer to the President's, eventually won out in that debate. More recently, the hard-line view, supported by staunch archconservatives, again raised its head in the debate over how to conduct the strategic arms reduction talks with Moscow. The unceremonial dismissal of Eugene Rostow as the head of the Arms Control and Disarmament Agency merely verified this lopsided sway held by the hard-liners over the more moderate professionals in their mutual competition for the President's ear. If a well-rounded foreign policy, true to its name, should not be unduly influenced by any one group of area specialists or one particular point of view, the anomalous situation is not likely to change in the prevailing highly ideologically charged atmosphere.

• Another problem is more general, and not confined to Asia. When the President prides himself as the world's leader in an anti-Soviet crusade, that commitment may have three consequences: (1) It can easily become the subjective criterion for judging the rightness or wrongness of the foreign policy of our allies and friends. Relations with allies may suffer at times, when their interpretations of Soviet intentions differ from Washington's, as happened in 1982 with Britain and France over the gas pipeline deals with Moscow until Reagan lifted the ban. (2) The "usefulness" of a country to Washington's running battle with the Soviet Union will then determine the nature of its relations with the United States, often to the neglect of that country's internal policies and ideological commitment. As long as it is at odds with the Soviet Union, Communist China has the potential of eclipsing industrial Japan in military importance, and is certain to outshine democratic but neutral India. For the same reason, Pakistan is destined to be more "useful" to Washington than is India. On the same scales, Britain weighs heavier than Argentina, as it did during the Falkland crisis, even if all other factors are left out. (3) The same can be said of a country's "vulnerability" to perceived threats of Soviet bloc subversion, which makes the country a "domino" in the East–West conflict. A noncommunist authoritarian system such as El Salvador, despite its internal liabilities, would qualify for large amounts of U.S. military and economic aid in the face of the communist threat.[25] If the same logic is followed consistently, Honduras, threatened by the neighboring Sandinist Nicaragua, would deserve greater U.S. support than would Costa Rica, the oldest democracy in Latin America, if only because the latter is not perceived to be equally vulnerable to external threat thus far. No wonder the English-speaking members of the Caribbean community, such as Guyana and Jamaica, which "unfortunately" do not share the same Salvador-type vulnerability, are complaining that they are being left out from the Reagan "Caribbean initiative."

The toning down in 1983 of the earlier rhetoric that portrayed Central America as the dominoes of U.S.–Soviet conflicts does not seem to serve the Administration's image well: It is seen either to have had poor judgment before or as being inconsistent now. Paradoxically, it is the product of a policy which has lost its balance because of an overplay of national security, if not an outright unwise manipulation of the Soviet specter.

Two concerns underlie this discussion. First, when so much stake is put on combating the Soviet Union, success or failure of U.S. foreign policy as a whole will almost invariably be determined by the success or failure of that campaign alone. Since any concession to the Soviets in the course of diplomatic give and take could be interpreted as failure or even betrayal, it makes flexibility very difficult. When necessary compromises have to be made, as is often the case, such as over the Siberian pipeline embargo, the President is unduly vulnerable to attacks from critics as being inconsistent or "Carter-like."[26] Second, the disproportionate emphasis on winning the fight with the Soviets cannot but distort the priorities and lead to the kind of policy imbalances and paradoxical consequences already noted. In all foreign policy areas, only U.S. relations with Israel seem to be not equally affected by the Administration's Soviet obsession. Even there, however, they have to contend with the parallel interest of building a strategic consensus among moderate Arab states. Furthermore, the Reagan plan, which erroneously expects the Saudis to pave the way for a Lebanese–Israeli agreement and to forego the idea of an independent Palestinian state, is premised on the mistaken assumption that the Saudis share such fear and dread of Soviet Communism (as well as Arab radicalism) that they would back Washington on all these critical issues.[27] The imbalance in policy and perception created by the overblown Soviet scare in Washington is as alarming as it is incredible. Even the President's own Scowcroft Commission had to find that the alleged "window of vulnerability" was nonexistent, openly contradicting Reagan's long-held belief.

The imbalance actually started with the second half of the Carter Administration following the Afghan invasion in late 1979, but has been carried to its logical extension under Reagan, without any more redeeming results than under Carter.[28] The chapters in this volume have suggested, sometimes implicitly, remedies for many of the policy dilemmas just reviewed. Other lessons are self-apparent in our review in this concluding chapter. While they are not to be repeated here, the following suggestions bear further amplification for the sake of a sounder and more balanced understanding of how U.S. security interests abroad should be managed.

First, in its assumptions, Washington should cease to expect our allies to share with it the same kind of assessments of the Soviet Union, much less that out of a common fear and distaste for the Soviet brand of communism they will back us to the hilt on all major issues. Lowering the expectation will lower the leverage that many of our allies and friends now seem to wield over us, and can in turn increase our relative leverage, such as in dealing with Japan. Instead of appealing to Tokyo's sense of a common destiny with us in curbing Soviet influence, we should frankly remind it of the absolute importance of the sea lanes for Japan's own survival. To fuel the furnaces of Japan's factories, there is a tanker every 50 miles or so on the high seas, hauling oil and other raw materials around the clock from the Middle East to Yokohama. The maintenance of peace and security to keep the sea lanes open for these tankers now depends on the power, at immense cost, of the United States.[29] If Washington pleads with the Japanese to help the United States fight a dreaded enemy, they would be doing us a big favor if they increase one penny in their defense budget. But, if we are not so overwhelmingly concerned with the alleged common enemy, the defense of the sea lanes and the preservation of peace on the high seas would worry the Japanese more than they do now. They would then realize that they will be doing themselves a favor by assuming greater defense responsibilities, quite independently of our urgings. The Japanese would then plead with us to do them the favor of sharing part of their security costs. They would then worry that U.S. commitments are not sufficient or forthcoming. What is true with Japan is true in many other cases, including the U.S. courting of China in the anti-Soviet game.

Second, still better, Washington should adjust its concerns about the Soviet Union to fit the reality. The reality today is that there are two, not one, superpowers, and that fact is not going to be changed, no matter what the United States may think or do. The worst that could happen is to play a "zero sum game," denying the Soviets their claim to superpower status and driving them into a game of revengeful reciprocation (or double defection). The balance of terror, in the era of bisuperpower nuclear parity, has rendered the traditional U.S. concept of unilaterally relying on its own preponderant power totally obsolete and operationally dangerous. Instead of escalating the contention and distrust, each superpower owes it to itself and humankind to strive toward changing U.S.–Soviet relationships from the "prisoner's dilemma" (double defection) to a "game of the chicken" (double cooperation). We should abandon the mentality of supply-side deterrence, return to MAD and reliance on a credible second-strike capability, and above all show willingness to accept a Soviet role in the world affairs of the 1980s and beyond. We can help instill a sense of responsibility in the Soviets as the other superpower by bringing them into the quest for solutions to the world's problems, including peace and scarcity. We can do so only if our relations with the Soviet Union are not a matter of pure military competition. If would-be delinquents can be turned into good cops in the domestic society, there is much that can be said about the prospect of turning the Soviets into a more responsible actor on the world scene.

Third, by the same token, we should not try to play the Chinese against the Soviets under the assumption that it would serve U.S. national interests. Instead, we should be prepared to accept a reasonable amelioration of Sino–Soviet relations. It has been argued elsewhere that the United States actually has a great deal to gain from such an amelioration. In the three-person game, the United States can play the role of the pivot only if (1) both of the two wing players, China and the Soviet Union, are concerned about remaining in good grace with the pivot, (2) either wing player is concerned lest the other move closer to the pivot at its own expense, and (3) both wing players have some minimal "common ground" between them, so that (1) and (2) can be present.[30] Continuing hostility between China and the Soviet Union, that is, absence of (3) in this light, actually leaves the United States, the pivot, very little leverage over either, on account of the absence of conditions (1) and (2). Playing the China card, which grew out of Washington's anti-Soviet impulse and a desperate concern about Soviet nuclear superiority, therefore, does not necessarily serve U.S. security interests, as is generally assumed. Instead, we should not only return to the pursuit of even-handedness in our relations with the two communist giants, but should encourage them to develop a reasonable rapport between them, which will actually enhance the United States' hand as the pivot power.

Last, but not least, the most meaningful lesson that can be drawn from the studies in this volume is that national security is far from military defense. For example, crucial measures just mentioned, such as turning the table on Japan, converting the U.S.–Soviet collision course into one of mutual statesman-like responsibility for world affairs, and maintaining the United States as the pivot power by promoting some "common ground" between Moscow and Peking, are political engineering feats. They require an informed and balanced judgment, depth of knowledge, vision, and imaginativeness that can hardly be expected from a pure military strategist or an area specialist, much less from one who has an axe to grind about the Soviet system and attempts to use the power at Washington's disposal to bring about its transformation or downfall.[31]

National security is too serious a matter to be left to our security managers alone. That is the conviction that has generated the chapters that make up this volume. Although it may sound like a cliché, our principal finding is that worrying about national security single-mindedly and in isolation, as though with blinders on, is not the wisest way of managing our national interests.

Traditionally, national security and foreign policy may be in a competitive relationship in the United States, in both their concerns and lines of accountability. In terms of its concerns, foreign policy has to reflect the interests of the American public, such as trade, travel, investment, and missionary evangelism abroad. Since it is subject to the American democratic process, foreign policy has to operate within the constraints of the habitual conflict of domestic interests and the public's natural aversion for maintaining a huge defense budget in peacetime. In normal times, the public is opposed to the use of economic embargo in pursuit of foreign-policy goals; and this certainly goes against the government's national security impulse, as was demonstrated in Carter's and Reagan's economic sanctions against Moscow. Despite U.S. involvement in World War I (which in itself was prompted by German sinking of the *Lusitania*, and by the desire to go to a "war to end all wars"), the postwar U.S. public's impulse was to return to normalcy and, in foreign policy, to push for disarmament and banning of war as an instrument of policy (hence, the Kellog–Briand Pact of 1928). Similarly, after World War II and the Korean War, the public wanted immediate demobilization and a speedy return to normalcy. A mounting public weariness contributed to the nonwinning of the Vietnam War and impelled Washington to end the war short of its earlier postulated goals and to bring the boys home. To the American public, order is the norm. In the U.S. tradition, there is a Lockian trust in rationality, which places an enormous emphasis on discussion (negotiation) and compromise. There is, in addition, an aversion to violence and a belief that war should be banned, except in collective defense against armed aggression and in a "just war," however defined.[32]

All these are at variance with the concerns and process of national security. The process of national security planning is a private reserve of the elite, or the decision makers, justified by the need for secrecy. It is therefore often beyond the routine lines of accountability. The words "security" and "secrecy" are in this connection synonymous. In terms of its concerns, national security follows a different impulse, toward the building of a greater defense force and spending more on military hardware. It has to be prepared for the worst. It is constantly keen on the possible "uses" of military force, such as: its deterrent role; its compellent role; the acquisitive use of force; intervention; counterintervention; collective action; and providing the necessary backdrop for diplomacy.[33] Preoccupation with the uses of military force may, paradoxically, necessitate endeavors (including an escalating cut-throat arms race) that actually go against the promotion of the requisite conditions for peace. The policy paradox reviewed in this chapter is, therefore, not an isolated problem, certainly not new with the Reagan Administration.

There are, in fact, many points of convergence between foreign policy and national security concerns. Both share (1) an American utopianism (ideals of Christianity and the philosophy of Enlightenment), (2) a sense of mission (translated into programs such as the "Good Neighbor Policy" of FDR, the Marshall Plan, and the human rights policy), (3) a crusading spirit ("to make the world safe for democracy" during World War I or "to protect democracy abroad" in Vietnam under Lyndon B. Johnson), and (4) a deep-rooted hatred for Soviet bloc communism.

In the hands of an unscrupulous President, the supremacy of national security might conveniently turn into a justification why he, the Commander-in-Chief, is above accountability, as indeed happened during Watergate. Even under less trying circumstances, a President skillful at maximizing the points of convergence just noted can do much to push for a particular line of policy consistent with his own idiosyncratic views on security, until he is stopped by Congress and the public. While the tensions between the two sets of concerns and the two processes are not the main thrust of our inquiry, their existence should not escape our attention, because it explains why the kind of paradoxes we have noted in the Reagan record came into being. While the problem is not new, the Reagan Administration's hypersensitivity about security and its penchant for reducing every foreign policy issue into a security issue have only magnified the severity of the problem. This is why we have stated that to be obsessed with national security is not necessarily the wisest way to manage our security interests. We believe that security concerns should be balanced with broad foreign policy interests, and the process of national security planning should be fully integrated with the foreign policy process. Only then will national security be freed from the idiosyncracies of a particular Presidential team, be made more open to the Constitutional lines of accountability, and have the potential of eschewing the kind of paradoxes that mark the Reagan record under review.

NOTES

1. Cf. Robert E. Osgood, "The Revitalization of Containment," *Foreign Affairs*, Vol. 60, No. 3 (Winter 1981), pp. 465–502.

2. Secretary Haig's speech before the American Society of Newspaper Editors in Washington, D.C., April 21, 1981; also "A Strategic Approach to American Foreign Policy," an address before the American Bar Association, in New Orleans, August 11, 1981, Depart-

ment of State, *Current Policy*, No. 305 (August 11, 1981); Secretary Shultz, "U.S. Foreign Policy: Realism and Progress," address before the 37th United Nations General Assembly, September 30, 1982, Dept. of State, *Current Policy*, No. 420 (September 30, 1982).

3. *New York Times*, January 20, 1983.
4. *Ibid.*
5. Secretary Shultz, before Senate Foreign Relations Committee, September 10, 1982, Dept. of State, *Current Policy*, No. 418 (September 10, 1982), p. 2.
6. Merely criticizing Begin is not going to produce results. Cf. article by former Presidents Ford and Carter, "A Time for Courage in the Middle East," in *Reader's Digest*, February 1983.
7. Alan Riding, "Violence Rules Central America Despite Pacts and Plans for Peace," *New York Times*, January 23, 1983, p. 1.
8. *U.S. News & World Report*, January 24, 1983, p. 11.
9. *Ibid.*
10. *Ibid.*
11. *New York Times*, January 5, 1983.
12. President Reagan's State of the Union message suggested a $55 billion defense budget trim for the next 5 years, which includes the $8 billion cut already noted, *New York Times*, January 26, 1983.
13. Flora Lewis, "Tacks and Ticks," *New York Times*, August 17, 1982.
14. Colin S. Gray, *The Geopolitics of the Nuclear Era* (New York: Crane, Russak, 1977), p. 66.
15. James C. Hsiung and Winberg Chai, eds., *Asia and U.S. Foreign Policy* (New York: Praeger, 1981), pp. 117, 231.
16. Thomas Schelling, *Strategy of Conflict* (New York: Oxford University Press, 1960), esp. p. 195f.
17. John Von Neumann and Oskar Morgenstern, *Theory of Games and Economic Behavior* (Princeton, N.J.: Princeton University Press, 1964); A. Rapaport and A. Chammah, *The Prisoner's Dilemma* (Ann Arbor: University of Michigan Press, 1970); Karl W. Deutsch, *The Analysis of International Relations*, 2nd ed. (Englewood Cliffs, N.J.: Prentice-Hall, 1978), pp. 145-149. Also see Patrick Morgan, *Deterrence: A Conceptual Analysis* (Beverly Hills, Calif.: Sage Publications, 1977), and Richard Rosecrance (ed.), *The Future of the Strategic System* (San Francisco: Chandler, 1972).
18. Herbert Scoville, Jr., "The MX Invites Attack," *New York Times*, Op Ed., December 13, 1982. Scoville is former Assistant Director of the Arms Control and Disarmament Agency and Deputy Director for Research for the CIA.
19. U.S. Dept. of State, "Communist Interference in El Salvador," *Special Report*, No. 80 (February 23, 1981).
20. IISS, *The Strategic Survey, 1981–1982* (London, 1982), p. 30.
21. See John Garver's chapter in this volume (Chapter 5), reference for Figure 1, for a discussion of these straits that link the Pacific and Indian Oceans.
22. I am grateful for this knowledge to Admiral Tun-hua Ke (retired), member of the Society for Strategic Studies (Republic of China), Taipei.
23. The Andropov counterproposal calls for reducing the Soviet intermediate-range missiles targeted at Western Europe from over 600 to 162, the combined total of the British and the French missiles in place, in exchange for the scrapping of the projected installation of U.S. Pershing II and ground launched cruise missiles in Western Europe at the end of 1983.
24. In addition to Hamrin's chapter in this book, see also A. Doak Barnett, *China's International Posture: Signs of Change* (Washington, D.C.: Asia Society China Council, 1982).
25. President Reagan certified "progress" in El Salvador, despite its "great obstacles," on human rights, on January 21, 1983, qualifying that country for $25 million U.S. military aid and possibly more for the 1983 fiscal year. *New York Times*, January 22, 1983.

26. Tom Wicker, "Looking Like Jimmy," *New York Times*, November 19, 1982, p. A35.

27. Thomas L. Friedman, "Pique in Washington over Those 'Moderate' Saudis," *New York Times*, January 23, 1983, p. E2.

28. See a critique of the Carter policy in Hsiung and Chai, *Asia and U.S. Foreign Policy*, note 15, pp. 240–245.

29. James Reston, "The Nakasone Talks," *New York Times*, January 19, 1983.

30. Gerald Segal, "China and the Great Power Triangle," *China Quarterly*, No. 83 (September 1980), p. 501; Lowell Ditmmer, "The Strategic Triangle: An Elementary Game-Theoretical Analysis," *World Politics*, Vol. 33, No. 4 (July 1981), pp. 485–515; James C. Hsiung, "China's Security Strategy, Anti-Hegemonism, and U.S. Policy: Analyses from a Three-Person-Game Perspective," paper delivered at the American Political Science Assocation annual meeting, September 1981, in New York. This study is being expanded into a forthcoming book.

31. On the impracticality of any such attempt to bring about internal transformation of the Soviet system from the outside, see Serwyn Bialer and Joan Afferica, "Reagan and Russia," *Foreign Affairs*, Vol. 61, No. 2 (Winter 1982), pp. 249–271.

32. Amos A. Jordan and William J. Taylor, Jr., *American National Security: Policy and Process* (Baltimore: Johns Hopkins Univ. Press, 1981), p. 51.

33. *Ibid.*, pp. 28–29.

Bibliography

DOCUMENTARY SOURCES

Foreign Broadcast Information Service. Various Issues.

International Monetary Fund. (IMF) *Direction of Trade Yearbook, 1980–81.* Washington, D.C. 1981.

U.S. Congress. House Committee on Foreign Affairs. Subcommittee on Asian Pacific Affairs. *The New Era in East Asia.* Washington, D.C.: U.S. Government Printing Office, 1981.

———. Committee on Foreign Affairs. Subcommittee on Asian and Pacific Affairs. *The United States and the People's Republic of China: Issues for the 1980's.* Washington, D.C.: U.S. Government Printing Office, 1980.

———. Committee on Foreign Affairs. Subcommittee on Asian and Pacific Affairs. *Playing the China Card: Implications for the United States–Soviet–Chinese Relations.* Washington, D.C.: U.S. Government Printing Office, 1979.

———. Senate. Committee on Foreign Relations. *Sino–American Relations: A New Turn.* Washington, D.C.: U.S. Government Printing Office, 1979.

———. Senate. Committee on Foreign Relations. *The Implications of U.S.–China Military Cooperation.* Washington, D.C.: U.S. Government Printing Office, 1982.

U.S. Department of Commerce. Bureau of the Census. *U.S. Statistical Abstract.* Washington, D.C.

U.S. Department of Defense. *Annual Report.*

U.S. Department of Interior. U.S. Bureau of Mines. *Mineral Yearbook. 1980. Volume III. Area Reports: International.* Washington, D.C. 1982.

U.S. Department of State. *Department of State Bulletin.* Various Issues.

———. *Current Policy.* Various Issues.

———. *U.S.–China Security Relationship.* Gist. Washington, D.C., 1980.

U.S. Library of Congress. Congressional Research Service. *Soviet Strategic Objectives and SALT II: American Perceptions*. Washington, D.C.: Report No. 78-119F, 1978.

———. *Increased U.S. Military Sales to China: Arguments and Alternatives*. Washington, D.C., May 20, 1981.

———. *China–U.S. Relations*. Issue Brief IB 76053, Periodically Updated.

ARTICLES

Bialer, S., and J. Afferica. "Reagan and Russia" *Foreign Affairs*, Vol. 61, No. 2, Winter 1982.

Bundy, M. et al. "Nuclear Weapons and Atlantic Alliance" *Foreign Affairs*, Vol. 60, No. 4, Spring 1982.

Burt, R. "The Evolution of the U.S. START Approach." *NATO Review*, Vol. 30, No. 4, 1982.

Chanda, N. "A U.N. Dove Flies into Turbulence." *Far East Economic Review*, March 27, 1981.

Church, F. "America's New Foreign Policy." *New York Times Magazine*, August 23, 1981.

Curtis, M. "Introduction After the Withdrawal from Sinai." *Middle East Review*, Vol. XIV, Nos. 3–4, 1982.

Ditmmer, L. "The Strategic Triangle: An Elementary Game-Theoretical Analysis." *World Politics*, Vol. 33, No. 4, July 1981.

Fallows, J. "The Great Defense Deception." *New York Review of Books*, May 28, 1981.

Garret, B. "The China Card: To Play or not to Play." *Contemporary China*, Spring 1979.

Hyland, W. "Avoiding a Showdown." *Foreign Policy*, No. 49, Winter 1982–1983.

Jacobsen, C. G. "Soviet-American Policy: New Strategic Uncertainties. *Current History*, October 1982.

Komer, R. W. "Maritime Strategy vs. Coalition Defense." *Foreign Affairs*, Vol. 60, No. 5, Summer 1982.

Lewis, F. "Tacks and Ticks." *New York Times*, August 17, 1982.

Bibliography / 195

Lewis, K. "The Reagan Defense Budget: Prospects and Pressures." The Rand Corporation P-6721, December 1981.

Monfort, C. A. "An MX 'Dense Pack' Would Need ABMs, Both Periling Security." *New York Times*, Op Ed page, December 1, 1982.

Niehaus, M. "Southeast Asia." *Chronologies of Major Developments in Selected Areas of Foreign Affairs*. Washington Committee Print Cumulative Edition, 1981.

Ocampo, S. "Guns are Not the Only Answer." *Far East Economic Review*, May 8, 1981.

Osgood, R. "The Revitalization of Containment." *Foreign Affairs*, Vol. 60, No. 3, Winter 1981.

Podhoretz, N. "The Neo-Conservative Anguish over Reagan's Foreign Policy." *New York Times Magazine*, May 2, 1982.

Porter, G. "The 'China Card' and U.S. Indochina Policy." *Indochina Issues*, Washington, D.C. Center for International Policy No. 11, 1980.

Reston, J. "The Tragedy of Begin." *New York Times*, September 22, 1982.

——. "The Nakasone Talks." *New York Times*, November 19, 1982.

Rogers, B. W. "The Atlantic Alliance." *Foreign Affairs*, Vol. 60, No. 5, Summer 1982.

Saunders, H. H. "An Israeli Palestinian Peace." *Foreign Affairs*, Vol. 61, No. 1, 1982.

Scotville, H., Jr. "The MX Invites Attack." *New York Times*, December 15, 1982.

Segal, G. "China and the Great Power Triangle." *China Quarterly*, No. 83, September 1980.

Smith, H. "How Many Billions for Defense." *New York Times Magazine*, November 1, 1981.

Stubbing, R. "The Imaginary Defense Gap: We Already Outspend Them." *The Washington Post*, February 15, 1982.

Wicker, T. "Looking Like Jimmy," *New York Times*, November 19, 1982.

Young, P. L. "China's Military Capabilities." *Asian Defense Journal*, February 1979.

Zasloff, J. J., and M. Brown. "The Passion of Kampuchea." *Problems of Communism*, January/February 1979.

BOOKS

Alexander, Y. and A. Nanes (ed.) *The United States and Iran: A Documentary History*. Frederick, Md.: University Publications of America, 1980.

Arnold, A. *Afghanistan. The Soviet Intervention in Perspective*. Stanford, Calif.: Hoover Institution Press, 1981.

Atlantic Council's Working Group on Security Affairs et al. *After Afghanistan: The Long Haul Safeguarding Security and Independence in the Third World*. Boulder, Colo.: Westview, 1980.

Ayoob, M. (ed.) *The Middle East in World Politics*. New York: St. Martin's Press, 1981.

Bain, K. R. *The March to Zion: United States' Policy and the Founding of Israel*. College Station: Texas A & M Univ. Press, 1979.

Barnett, A. D. *China's International Posture: Signs of Change*. Washington, D.C.: Asia Society China Council, 1982.

———. *U.S. Arms Sales: The China–Taiwan Tangle*. Washington, D.C.: The Brookings Institution, 1982.

———. *China and the Major Powers in East Asia*. Washington, D.C.: The Brookings Institution, 1977.

BDM Corporation. *U.S.–PRC–USSR Triangle: An Analysis of Options for Post-Mao China*. Vienna, Virginia, 1976.

Bradley, P. C. *The Camp David Peace Process: A Study of the Carter Administration Policies 1977–1980*. Crantham, N.H.: Tompson & Rutter, 1981.

Carter, J. *Keeping Faith*. New York: Bantam Books, 1982.

Chen, K. C. *China and the Three Worlds*. White Plains, N.Y.: M. E. Sharpe, 1979.

Chiu, H. *China and the Taiwan Issue*. 2nd ed. New York: Praeger, 1979.

Clough, R. N. *East Asia and U.S. Security*. Washington, D.C.: The Brookings Institution, 1975.

———. *Island China*. Cambridge, Mass.: Harvard Univ. Press, 1978.

Cohen, W. I. *America's Response to China*. 2nd ed. New York: John Wiley, 1980.

Collins, J. M. *U.S. Soviet Military Balance, 1960–1980*. New York: McGraw-Hill, 1980.

Congressional Quarterly Inc. *China: U.S. Policy Since 1945*. Washington, D.C.: Congressional Quarterly, 1980.

———. *The Middle East*. 5th ed. Washington, D.C.: Congressional Quarterly, 1981.

Ellison, R. *Sino–Soviet Relations: A Global Perspective*. Seattle, Wash.: Univ. of Washington Press, 1981.

Epstein, W. *The Last Chance: Nuclear Proliferation and Arms Control*. New York: Free Press, 1976.

Eveland, W. *Ropes of Sand: America's Failure in the Middle East*. New York: W. W. Norton, 1980.

Fallows, J. M. *National Defense*. New York: Random House, 1981.

Fraser, T. G. (ed.). *The Middle East, 1914–1979*. New York: St. Martin's Press, 1980.

Gaddis, J. L. *Strategies of Containment: A Critical Appraisal of Postwar American National Security Policy*. New York: Oxford Univ. Press, 1982.

Garver, J. W. *China's Decision for Rapprochment With the United States, 1968–1971*. Boulder, Colo.: Westview Press, 1982.

George, J. L. *Problems of Sea Power as We Approach the Twenty-first Century*. Washington, D.C.: American Enterprise Institute, 1978.

Gilbert, S. P. *Northeast Asia in U.S. Foreign Policy*. Beverly Hills, Calif.: Sage, 1979.

Goodman, G. K., and F. Moss (eds.). *The United States and Japan in Western Pacific*. Boulder, Colo.: Westview, 1980.

Gottlieb, T. *Chinese Foreign Policy Factionalism and the Origins of the Strategic Triangle*. Santa Monica, Calif.: Rand R-1902-NA. November 1977.

Gray, C. S. *The Geopolitics of the Nuclear Era*. New York: Crane, Russak, 1977.

———. *Strategic Policies and Public Policy: The American Experience*. Lexington, Ky.: University Press of Kentucky, 1982.

Harding, H. *Evolving Strategic Realities: Implications for U.S. Policymakers*. Washington, D.C.: National Defense University Press, 1980.

Harrison, S. S. *The Widening Gulf: Asian Nationalism and American Policy*. New York: Free Press, 1978.

Hinton, H. C. *Three and a Half Powers: The New Balance in Asia.* Bloomington, Ind.: Indiana University Press, 1975.

Hollick, A. L. *United States Foreign Policy and the Law of the Sea.* Princeton, N.J.: Princeton University Press, 1981.

Hsiung, J. C., and W. Chai (ed.). *Asia and U.S. Foreign Policy.* New York: Praeger, 1981.

Hsiung, J. C., and S. S. Kim (eds.). *China in the Global Community.* New York: Praeger, 1980.

Hsiung, J. C. *Law and Policy in China's Foreign Relations.* New York: Columbia Univ. Press, 1974.

Imai, R., and H. S. Rowen. *Nuclear Energy and Nuclear Proliferation: Japanese and American Views.* Boulder, Colo.: Westview, 1979.

International Institute for Strategic Studies (IISS). *Military Balance, 1980–1981.* London, 1981.

——. *Strategic Survey 1979.* London, 1980.

Jo, Y. *U.S. Foreign Policy in Asia: An Appraisal.* Santa Barbara, Calif.: ABC Clio, 1978.

Johnson, S. E. *The Military Equation in Northeast Asia.* Washington, D.C.: Brookings Institution, 1979.

Jordan, A. A., and W. J. Taylor, Jr. *American National Security: Policy and Process.* Baltimore, Md.: Johns Hopkins Univ. Press, 1981.

Jorgensen-Dahl, A. *Regional Organisation and Order in Southeast Asia.* London: Macmillan, 1982.

Kenen, I. L. *Israel's Defense Line: Her Friends and Foes in Washington.* Buffalo, N.Y.: Prometheus Books, 1981.

Kim, S. S. (ed.). *Documents on Korean–American Relations: 1943–1976.* Seoul: Research Center for Peace and Unification, 1976.

Kim, S. S. *China, the United Nations, and World Order.* Princeton, N.J.: Princeton Univ. Press, 1979.

Kim, Y. H. *American Frontier Activities in Asia: United States Asian Relations in the Twentieth Century.* Chicago: Nelson Hall, 1981.

LaFeber, W. *America, Russia and the Cold War*. New York: John Wiley, 1980.

Ledeen, M. A. *Debacle, The American Failure in Iran*. New York: Knopf, 1981.

Lieberthal, K. G. *Sino–Soviet Conflict in the 1970s: Its Relations and Implifications for the Strategic Triangle*. Santa Monica: Calif.: RAND Corporation, 1978.

Lewy, G. *America in Vietnam*. New York: Oxford Univ. Press, 1980.

Mackinder, H. *Democratic Ideals and Reality*. New York: Norton, 1962.

Maghroori, R. *The Yom Kippur War: A Case Study in Crisis Decision Making in American Foreign Policy*. Washington, D.C.: University Press of America, 1981.

Mahan, A. *The Influence of Sea Power on History*. London: Sampson Low, Marton & Co., 1892.

Meyer, G. E. *Egypt and the United States: The Formative Years*. Rutherford, N.J.: Farleigh Dickinson Univ. Press, 1980.

Morgan, P. *Deterrence: A Conceptual Analysis*. Beverly Hills, Calif.: Sage Publications, 1977.

Morrison, C. E. and A. Suhrke. *Strategies of Survival: The Foreign Policy Dilemmas of Smaller Asian States*. New York: St. Martin's Press, 1979.

Myers, R. H. (ed.). *Two Chinese States: United States' Foreign Policy and Interests*. Stanford, Calif.: Hoover Institution Press, 1978.

Newhouse, J. *Cold Dawn, the Story of SALT*. New York: Holt, Rinehart & Winston, 1973.

Noyes, J. H. *The Clouded Lens: Persian Gulf Security and U.S. Policy*. Stanford, Calif.: Hoover Institution Press, 1982.

Oksenberg, M., and R. B. Oxnam (eds.). *Dragon and Eagle: United States-China Relations: Past and Future*. New York: Basic Books, 1978.

Pandley, B. N. *South and Southeast Asia 1945–1979: Problems and Policies*. London: Macmillan, 1980.

Pfaltzgraff, R. L., Jr. *Energy Issues and Alliance Relations: The United States, Western Europe, and Japan*. Cambridge, Mass.: Institute of Foreign Policy Analysis, 1980.

Poole, P. A. *Eight Presidents and Indochina* Rev. ed. Huntington, N.Y.: Robert E. Krieger, 1978.

Pringle, R. *Indonesia and the Philippines: American Interests in Island Southeast Asia*. New York: Columbia Univ. Press, 1980.

Quandt, W., F. Jabber, and A. M. Lesch. *The Politics of Palestinian Nationalism*. Berkeley: Univ. of California Press, 1973.

Rosecrance, R. (ed.). *The Future of Strategic Systems*. San Francisco: Chandler, 1972.

Rubin, B. *Paved with Good Intentions: The American Experience and Iran*. New York: Oxford Univ. Press, 1980.

Rudolph, L. and S. Rudolph. *The Regional Imperative: The Administration of U.S. Policy Towards the South Asian States under Presidents Johnson and Nixon*. Atlantic Highlands, N.J.: Humanities, 1980.

Scalopino, R. *Asia and the Road Ahead: Issues for the Major Powers*. Berkeley: Univ. of California Press, 1975.

——. *Economic, Political, and Security Issues in Southeast Asia in the 1980's*. Berkeley: Univ. of California Press, 1982.

——. *The United States and Korea: Looking Ahead*. Beverly Hills, Calif.: Sage, 1979.

Schelling T. *Strategy of Conflict*. New York: Oxford Univ. Press, 1960.

Simon, S. W. *The ASEAN States and Regional Security*. Stanford, Calif.: Hoover Institution Press, 1982.

Solomon, R. H. *Asian Security in 1980's: Problems and Policies for a Time of Transition*. Santa Monica, Calif.: RAND Corporation, 1979.

Spykman, N. *Geography of the Peace*. New York: Harcourt and Brace, 1944.

Stoessinger, J. *Nations in Darkness: China, Russia and America*. 3rd ed. New York: Random House, 1978.

Stuart, D. T., and W. T. Tow (eds.). *China, The Soviet Union, and The West*. Boulder, Colo.: Westview, 1982.

Swearington, R. *The Soviet Union and Postwar Japan, Escalating Challenge and Response*. Stanford Calif.: Stanford Univ. Press, 1978.

Tahir-Kheli, S. *The United States and Pakistan: The Evolution of an Influence Relationship*. New York: Praeger, 1982.

Tucker, R. W. *The Purposes of American Power: An Essay on National Security*. New York: Praeger, 1981.

Weinstein, F. B. *U.S.-Japan Relations and the Security of East Asia: The Next Decade.* Boulder, Colo.: Westview, 1978.

Weinstein, F. B., and K. Fuji (eds.). *The Security of Korea: U.S. and Japanese Perspectives in the Nineteen Eighties.* Boulder, Colo.: Westview, 1980.

Whiting, A. S. *The Chinese Calculus of Deterrence: India and Indochina.* Ann Arbor: Univ. of Michigan Press, 1975.

Wu, Y. *U.S. Policy and Strategic Interests in the Western Pacific.* New York: Crane-Russak, 1975.

Zagoria, D. S. (ed.). *Soviet Policy in East Asia.* New Haven, Conn.: Yale Univ. Press, 1982.

Zasloff, J. and M. Brown. *Communist Indochina and U.S. Foreign Policy: Forging New Relations.* Boulder, Colo.: Westview, 1978.

Index

A-10, 22
ABM (anti-ballistic missile systems), 4, 13, 179
Abramowitz, Morton, U.S. Ambassador to Thailand, 117
Afghanistan, 9, 16, 26, 34, 38, 53, 54, 65, 66, 67, 73, 131, 135, 147, 151, 153, 154, 156–157, 165, 176, 180,184
Africa, 7, 33, 176
  Horn of, 131
  North, 131
  South, 70, 160
  Southern, 11
ALCM (air-launched cruise missile), 5
Algeria, 131
America
  Central, 161, 184
  North, 80, 160
  South, 80
Andropov, Yuri, 153, 154, 155, 156, 157, 158, 176, 181
Angola, 33, 150, 176
ANZUS Conference, 103
Arab, United Arab Republic, 132
Arafat, Yasir, 139
arctic waters, 10
Argentina, 71, 183
Armacost, Michael H., 93
ASEAN, 69, 85, 86, 87, 90, 92, 93,
  mineral resources, 87–89
Asia, 15, 16, 19–20, 23, 27–29, 30, 45
  Central, 156
  collective security system, 54
  East, 9, 20, 24, 41
  Northeast, 14, 20, 23, 27, 28, 30, 45
  power balance, 41
  Southeast, 9, 85, 93
  Southwest, vi, 9, 13, 21

ASW (anti-submarine warfare), 89, 90
Atlantic, 1, 9
  North, 7

B–1 bomber, 4, 5, 13, 178
B–52 bomber, 5
Bangkok, 93
Barents Sea, 10
Begin, Menachem, 138, 139, 141, 142, 143, 144
Behrain, 140
Benson, Lucy Wilson, 136
Berring Sea, 180
Bialer, Seweryn, 151
Boeing 707 aircraft, 37
Bonn, 150
Brazil, 87
Brezhnev, Leonid, 4, 12, 67, 74, 153, 157
Britain, 103, 183
Brown, Harold, 38, 104, 109
Brzezinski, 37, 46, 107
Buckley, James, 118
Bundy, McGeorge, 12
Bush, George, 35, 67, 100

Cairo Conference 1964, 132
Cam Rahn Bay, 16, 192, 113
Camp David peace process, 134, 139, 141, 142, 143, 173
Canada, 160
Caribbeans, 7, 10, 170, 175–176, 180, 184
Carrington, Lord, 139
Carter, 2–4, 6–8, 10, 11, 13–14, 20, 26–27, 34, 37, 38, 99, 102, 134, 136, 142, 150, 184
Chadid, Chazli Ben of Algeria, 140
Cheysson, Clade, 139

204 / U.S.–ASIAN RELATIONS

Chiang, Ching-Kuo, 81
    Kai-shek, 75
Chile, 87
China, 15, 20, 25, 27
Chinese, 7, 15
Chun, President (Korea), 22
Civil Defense Program, 5
Clark Air Force base, 102
Clark, William, 11, 176
Communist Party
    China, 94
    Indonesia, 94
    Malaysia, 94
    Thai, 93
compellence, 178
Costa Rica, 183
Cuba, 10, 13, 147, 160, 164, 180
Czechoslovakia, 73

Dannag, 16, 92
Delta-class ships, 10
Deng, Xiaoping, 66, 67, 69, 70, 73, 74, 76–78, 80, 103, 108, 109, 172
detente, 20
deterrence, 14, 21, 177–179
Dutch, 74
Dutch East Indies, 85

Egypt, 131, 132
Eisenhower, 143
El Salvador, 176, 180, 183
Eurasian land mass, 7
Euro–Asian land mass, 14
Europe, 1, 8, 9, 13, 45–47, 170, 176, 177
    Central, 6
    East, 13
    Western, v, 4, 7, 8, 14, 15, 28, 54

F–165, 22
Fahd, Crown Prince (now King) of Saudi Arabia, 138, 142
Falklands, War, 71, 183
Far East, 10, 13, 15, 16, 40
Fez Conference, 139, 142
Fischer, Dean (U.S. Dept. Spokesman), 139

Florida Straits, 10
FMS (Foreign Military Sales), 23
Ford, President, 4, 37, 102, 142, 150
French, 73, 80

Gandhi, 165
Gaza Strip, 132, 141, 143, 144
Geng Biao, 70, 118
Germany, 155
    West, 80
global defense, fwi, 15
Golan Heights, 133, 139
Gong Kedak, 103
Guam, 25, 89, 102, 103, 171
Guatemala, 176
Gulf Cooperation Council (GCC), 137
Gulf of Aqaba, 132, 133
Gulf of Thailand, 103
Guyana, 184

Habash, George, 139
Haig, Alexander, 11, 34, 36, 39, 69, 103, 113, 114, 115, 116, 118, 122, 135, 171, 174, 180
Hanoi, 67, 77, 93
Hassan, King of Morocco, 139, 142
Hawatme, Naif, 139
heartland power, 14, 177
Helsinki accords, 150
Hokkaido, 15
Holbrooke, Richard, 108
Holdridge, John, 25, 114–118, 122
Honduras, 176, 183
Hong Kong, 39, 71, 80, 104, 119
horizontal escalation, 14, 170
Hu Yaobang, 64, 67, 70, 74, 76, 78, 80, 118
Hua Guofeng, 67
Hua-lien, 180
Huang Hua, 64, 65, 70, 74
Hussein, King of Jordan, 139

ICBMC (Inter-Continental Ballistic Missiles), 4, 5
Ikle, Fred, 7
India, 42, 49, 66, 79, 156, 165, 183
Indian Ocean, 8–9, 45, 89, 102, 180, 288

Indian-Pacific Ocean, 1, 7, 9, 10, 15, 20, 103, 170, 177
Indo-China, fwi, 16, 34, 53, 66, 77, 91, 95
Indonesia, 85, 87, 88, 89, 91, 92, 180
Iran, 9, 131, 142, 160, 165, 218, 253, 264
 Iraq War, 136
 Shah of, 26, 34, 50
Iraq, 131, 135, 140, 142, 174
Israel, 69, 131, 132, 142, 160
 fight with Egypt, 133
 invasion of Lebanon, 140, 143, 174
ISA (International Seabed Authoriity), 2

Jackson-Vanik Amendment, 150
Jamaica, 184
Japan, 7-10, 15, 20, 22, 23-26, 27, 44, 49, 51, 54, 71, 89, 160
 defense against USSR, 48
 defense forces, 15
 role in regional balance, 71
 textbook controversy, 71
Jaruzelski, Polish general, 2, 176
Jordan, 131, 141, 144
Jordan River, 132, 133

Kampuchea, 66, 77, 85, 91, 92, 120, 121, 173
Kang Xi emperor. 75
Karmal, Babrak regime, 156
Kelatan, 103
Kennan, George, 12
Kennedy, John F., 2
Khmer Rouge, 109, 117, 118, 120, 122, 123
Khomeini, 8, 165
Kim Il-sung, 70
Kissinger, 133, 181
Kitti Kachorn, 99
Korea, 9, 30, 53, 156,
 North, 1, 22, 30, 182
 South, 7, 8, 10, 15, 21, 22, 23, 25, 30, 70, 160, 170, 175, 182
 War, 1

Kriangsak, Thai Prime Minister, 98-9, 112
Kurile Islands, 15

Lafayette missiles, 89
Laos, 85, 91
Latin America, 10, 53, 185
Law of the Sea Treaty, 1
Lebanon, fwv, 132
 Israeli invasion of , 140, 143, 174
Lee Kuan Yew, 100
Lehman, Jr., John F., 14
Lenin, 66, 67, 73
Li Xiannian, 70
Libya, 135, 142, 160, 174
Lombok Straits, 89

M-1 tank, 178
M-55, 23
Mackinder, Sir Halfard, 7, 14
MAD (mutual assured destruction), 4, 13, 178
Maghreb states of North Africa, 131
Mahan, Thayer, 14
Malaccan Straits, 180
Malaysia, 85, 87, 88, 92, 93, 103
Mao Zedong, 76
Manila Treaty, 99
Marcos, Ferdinand. Philippine President, 100, 102
Marxism-Leninism, China comment on, 79-80, 152
MBFR (Mutual and Balanced Force Reduction), 6
McNamara, Robert S., 12
Mediterranean, 9
Mexico, 160-161
Middle East, fwi, 7, 9, 45, 47, 53, 80, 185
Miller, J. D. B., 131
Minsk, 9
Minuteman II Missile, 4
Mitre Corp., 137
Mondale, Vice President, 67, 98, 100
Mongolia, 156
Multiple-front scenario, 13
MX super missile, 4, 13, 178, 179

206 / U.S.-ASIAN RELATIONS

Nakasone, Jasuhisa, 178, 182
Nambia, 176
NATO, 1, 3, 6, 7, 10, 150, 176
Nassar, Gamal Abdel, 132
Negroponte, John, 121
Neo-Mahan strategy, 14
Netherlands, the, 66
New Economic Policy, 66
New International Economic Order, 63
New Zealand, 97, 98, 103
Nicaragua, 183
Nigeria, 160
Nixon, 2, 7, 33, 37, 142, 150, 181
Nixon-Kissinger, 150
North-South Conference, Cancun, Mexico, 63
North-South dialogue, 63
Northwest Cape, 103
nuclear deterrence, 4, 6, 170
   escalation, 6
   modernization, 4
nuclear power, 2-3, 13, 164
   advantages of, 3
   arms reduction, 4
   arsenals, fwi, 4
   nuclear attack, 5
   balance-imbalance, 3, 4, 6, 170, 178
nuclear powered submarines, 9
NUTS (Nuclear Utilization Theories)
   logic, 13, 179

Okhotsk sea, 180
Oman, 140
Ombai-Water Straits, 89
OPEC, 87
Ostopolitik, 150
Ottoman empire, 97

Pacific
   fleet, 9
   Northwest, 24
   Ocean, 25
   power, 1
   sea lanes, 9
   western, 9-10, 16, 22, 28, 89, 180

Pakistan, 42, 78, 131, 156, 179, 183
Palestine Liberation Organization, 132-139, 141, 151
Pao Vang, 118
Peking, 34, 36
Pentagon, 4, 28
People's Daily, 36, 68, 73, 75
Persian Gulf, vi, 7-9, 15, 21, 28, 45, 89, 90, 170, 174
Petropavlosk, 10
Philippines, the, 25, 85, 91, 93, 94, 100, 171
Phnom Pehn, 120, 121, 124
Pin Gap, 102
Poland, 2, 53, 66, 80, 148, 153, 154, 176, 180
Pyong yang, 15, 23

Qaddafi Muammar, 140
Qatar, 140
Qing court, 75

RDF (Rapid Deployment Force), 9, 13, 21, 27
Reagan, Ronald, vi, 1-4, 11, 12, 22
   Asian Policy, 2
Republican Party Platform, 22
Rimland, a-la Spykman, 7
Roger, William P., 133
Rogers, General Bernard W., 12
Romania, 79
Rostow, Eugene, 176, 183
Roy, J. Stapleton, 117

Sadat, Anwar el, 138, 139, 140
SALT (Strategic Arms Limitation Talks), 5
SALT II Treaty, 34, 151
Saudi Arabia, 131, 132, 160, 174
Sea of Japan, 180, 181
security of
   Japan, 24
   Northeast Asia, 19, 20
   Saudi Arabia, 135
   South Korea, 170
   Thailand, 93, 98, 109, 112, 125
   U.S., 2
   Western Europe, 8

Seelye, Ambassador Talcott W., 140
Senate Foreign Relations Committee, 36
Seventh Fleet, 89
Shultz, George, 11, 36
Siberian development, 157
simultaneity, doctrine of, 13
Sinai Peninsula, 132, 133, 134
Singapore, 85, 103
Sino–(North) Korean cooperation, 70
Sino–Japanese relations, 66
Sino–Mongolian frontiers, 54, 65
Sino–Soviet relations
 border, 15, 54, 64
 dispute in Asia, 49
 power, 33, 38, 44
 reconciliation, 41, 54
 relations, 37, 64, 67
 split, 15
 talk, 74, 78
 threat, 13
Sino–U.S. relations
 alliance against the USSR, 38, 39, 45
 cooperation, 67, 69, 107, 112
 cooperation against Vietnam, 126
 detente, 66
 joint communique, 54, 64, 65
 normalization, 66, 70, 150, 151
 relations, 36, 37, 41, 49, 50–52, 63, 64, 67, 69, 78
 reconciliation, 11
 security arrangement, 35
Sino–Vietnamese confrontation, 16
SLBM (submarine-launched missile force), 4, 28, 180, 289
Smith, Gerard, 12
socialist nationalism, 157
Somali, 63
Son Sann (Kampuchean Third Force), 113, 117, 175
South China Sea, 15, 16, 103, 180
South–South consultation, 63
Soviet Union, 2, 3, 5, 11, 12, 14, 15, 21, 29, 43
 army brigade, 10
 aggression, 12, 13, 21, 170
 containment policy toward China, 64–5
 expansion, 2, 12, 20, 21, 23, 25, 26, 29, 33, 43–44, 68, 69, 94
 forces in China, 51
 forces in Cuba, 151
 forces in the Okhotsk Sea, 10, 15, 179
 influence in Asia, 54
  in Poland, 150
  in Vietnam, 16, 65
 MIG–21, 40
 military power, 1, 7, 10, 14, 33
 navy, 9–10, 14, 25
 nuclear attack against China, 39
 Pacific fleet force, 9–10, 15, 16, 54, 180, 181
 ties with North Korea, 28
 relations with the U.S., 166
 improvement of relations with China, 156
Soya Strait, 15
St. Petersburg, 157
Stalin, 74, 153
START (Strategic Arms Reduction Talks), 5, 6, 152
Stinger, the, 22
Stoessel, Walter, 115
Straits of Hormuz, 135
Straits of Malacca, 89
Straits of Tiran, 132
Strategic Doctrine, 11
Su-ao, 180
Subic naval base in the Philippines, 102
Suez Canal, 132–33
Sunda Strait, 89
sustainability, 170
Syria, 131, 132, 135, 140, 174

Taiwan, 28, 33, 40, 42, 49, 67, 68–70, 80, 160, 164
Taiwan Strait, 180
Thailand, 42, 79, 85, 91, 92, 93, 140
Thai Communist Party (CPT), 93
Thai–Kampuchean border, 93, 109, 113, 118
Thai–Laos border, 93
three-ocean Navy, 169
Thatcher, Margaret, 71, 176

Third World, 13, 23, 33, 63, 68, 69, 154, 170
Tower, Senator John, 117
Trade Reform Act (of 1974), 150
Trident II submarine-launched missile force, 4, 89, 178
Truman, Harry, 2
Truong Nhn Tang, 117
Tsugrua Strait, 15
Tsuschima Strait, 15

United Nations, 2, 64, 108
United Nations Emergency Force (U.N.E.F.), 132
U.S. aid to
  El Salvador, 175–6
  Indonesia, 106
  Israel, 143–4
  Saudi Arabia, 135–7, 174
  South Korea, 170, 182
  Thailand, 104–106, 171
U.S. arms sales to
  China, 34, 38, 39, 41, 42, 47, 50, 52, 171
  Taiwan, 35, 36, 53, 54
U.S. interest in Southeast Asia, 89
U.S. investment in ASEAN, 87
  in Central America and Mexico, 87
  in Latin America, 87
U.S.–Kampuchean Emergency Group, 118
U.S. military forces, 6–9
U.S. military expansion, 170
U.S. policy on
  Asia, 1, 15, 20, 22
  Indochina refugees, 106–7, 123
  Southeast Asia, 92, 94
  Vietnam, 113–118, 173
U.S. relations with
  China and USSR, 45, 46, 52–3
  Israel, 164–5
  Japan, 71, 161
  South Korea, 27
  Soviet, 1, 33, 42
  Taiwan, 48, 50
  Pakistan, Turkey, Morocco, Egypt, Sudan, Somalia, Jordan, Oman, 136, 174

Vance, 37, 107, 150
Versailles, 69
Vietnam, 42, 49, 85, 92
  communists, 85
  conflict with neighbors, 165
  normalization with U.S., 107
  occupation of Kampucha, 92, 99, 104, 113, 114, 172
  regional expansionism, 172
  South, 91
  War, 1, 45, 54
  war with China, 66
Vladivostok, 9, 15, 25, 150, 180, 181
V/STOL aircraft carrier, 9

Walters, Vernon, 118
Warsaw Pact nations, 6
Washington, 1, 6, 7, 9, 16, 22, 24, 68
Weinberger, Caspar, 11, 13, 14, 24, 70, 102, 177
West Bank, 141, 143, 144
World War II, 1, 2, 13
Wu Xueqian, 78

Ye Jianying, 73
Yemen, 132
  North, 27
  South, 135, 174
Yokohama, 185
Yugoslavia, 80
Zhang Zhuzi, 118
Zhang Wenjin, 73
Zhao Ziyang, 63, 64, 65, 67, 78, 79
Zhou En-lai, 75, 76

About the Editor and Contributors

James C. Hsiung, Ph.D., Columbia University, is a Professor of Politics at New York University, where he teaches international politics, international law, and comparative politics. He is author of many books including *Law and Policy in China's Foreign Relations*. His latest works are: *China in the Global Community* (coedited with Samuel S. Kim), and *Asia and U.S. Foreign Policy* (coedited with Winberg Chai). He is an Executive Editor for *Asian Affairs*, a consulting editor for *World Affairs* quarterly, and President of the Contemporary U.S.–Asia Research Institute, Inc. He has been a consultant to a number of Asian governments, including the Ministry of Education, the Republic of Singapore.

Paul H. Borsuk, who maintains a professional and scholarly interest in the Soviet Union, did graduate work in international relations at the Johns Hopkins University School of Advanced International Studies and in Soviet studies at Columbia University. He served as an intelligence analyst with the U.S. Central Intelligence Agency from 1977 to 1981.

Winberg Chai, Ph.D., New York University, is a Distinguished Professor of International Studies and Humanities, University of South Dakota (Vermillion). He is author and coauthor of fifteen books on Chinese culture, history, and politics, including *The Search for a New China*. He coedited (with James C. Hsiung) *Asia and U.S. Foreign Policy*. He is a past president of the American Association for Chinese Studies, currently a special adviser to the Kingdom of Saudi Arabia, and listed in *Who's Who in the World*.

John W. Garver, who is on leave and doing field work in Asia, was on the Political Science Faculty of the University of Nevada (Reno). After receiving his Ph.D. from the University of Colorado in 1979, he taught at a number of universities and served as a political risk analyst in Washington, D.C. He is author of *China's Decision for Rapprochement with the United States, 1968–1971*, and a number of articles, the latest of which is "Sino–U.S. Relations and U.S. Arms Sales to Taiwan," *Orbis* (Winter 1982).

Carol Lee Hamrin, Ph.D., University of Wisconsin (Madison), is a research specialist on China, U.S. Department of State. Her interests include Chinese foreign policy strategy, Sino–U.S. relations, and U.S.–Soviet relations. In 1980–1982, on loan from the State Department, she was chief of the China Branch, Analysis Group, Foreign Broadcast Information Service (FBIS). Author of a number of papers on China, she was for sometime an adjunct professorial lecturer at the Johns Hopkins University School of Advanced International Studies.

Norman D. Levin, who is completing his Ph.D. dissertation on "The Evolution of Japanese Defense Strategy," for Columbia University, has since late 1979 been on the research staff of the Social Science Department, the RAND Corporation (Santa Monica, California). His main interests are Korea and Japan, and U.S. security policy in Northeast Asia.

Robert G. Sutter is an Asian affairs specialist with the Congressional Research Service. Trained as a historian (Ph.D., Harvard University) on modern China and modern Japan, he served a stint in 1968–1977 with the Central Intelligence Agency as an analyst on Chinese foreign policy. He is author of *China-Watch: Toward Sino-American Reconciliation* and *Chinese Foreign Policy after the Cultural Revolution, 1966–1977*, among others. He has taught courses on U.S.–East Asian relations at Johns Hopkins School for Advanced International Studies and at Georgetown University.